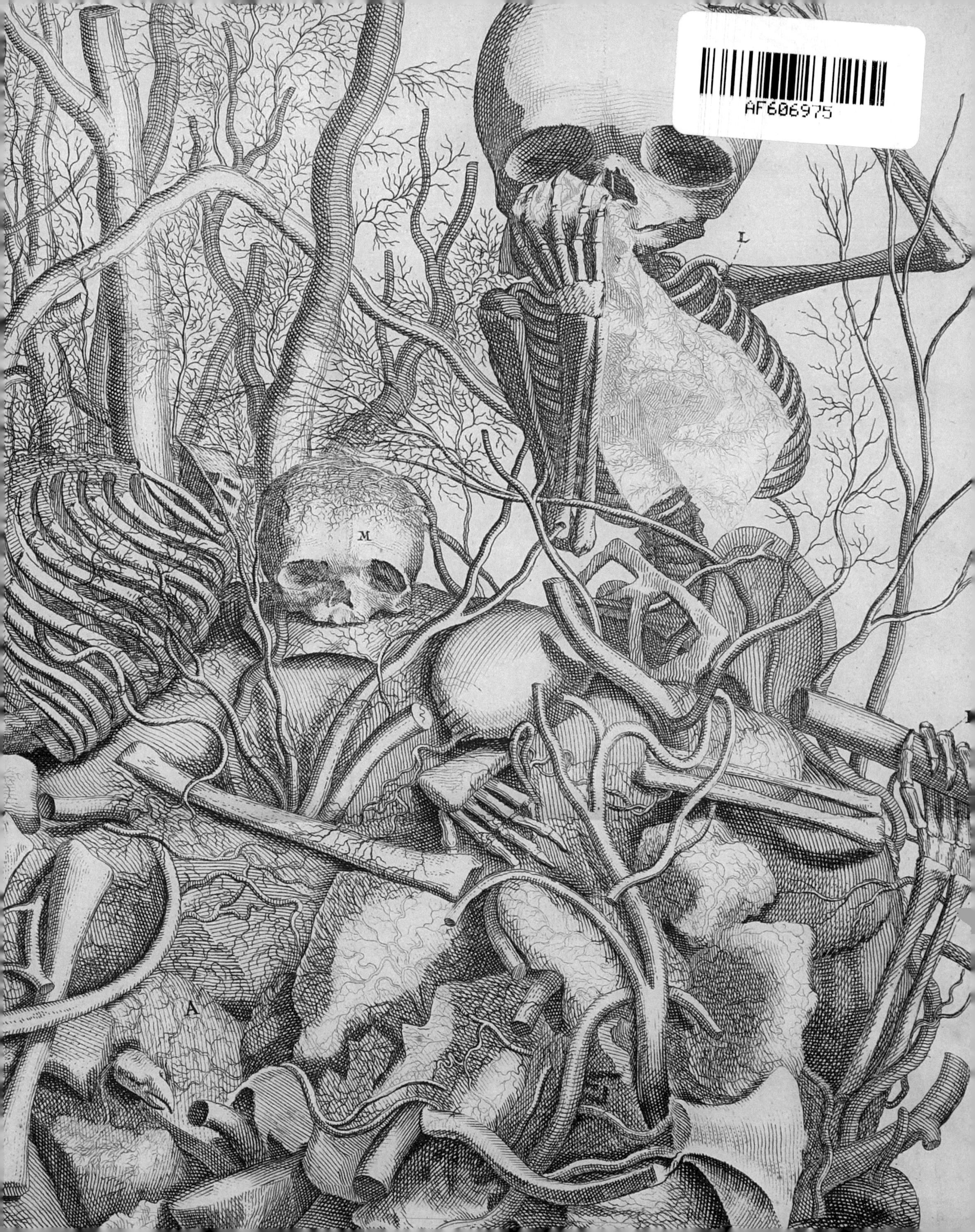
L
M
A

Frederik Ruysch

and His

Thesaurus Anatomicus

FREDERIK RUYSCH

AND HIS

THESAURUS ANATOMICUS

A Morbid Guide

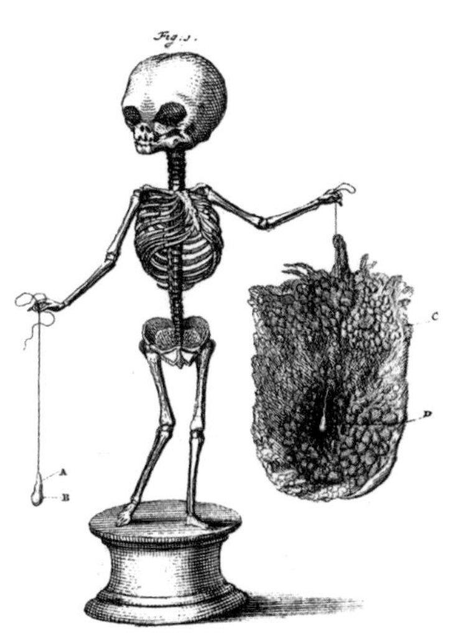

EDITED BY JOANNA EBENSTEIN

THE MIT PRESS, CAMBRIDGE, MASSACHUSETTS / LONDON, ENGLAND

The MIT Press would like to thank the anonymous peer reviewers who provided comments on drafts of this book. The generous work of academic experts is essential for establishing the authority and quality of our publications. We acknowledge with gratitude the contributions of these otherwise uncredited readers.

This book was set in Bembo, and designed by Joanna Ebenstein. Printed and bound in China.

Library of Congress Cataloging-in-Publication Data

Names: Ruysch, Frederik, 1638-1731, author. | Ebenstein, Joanna, 1971- editor.

Title: Frederik Ruysch and his thesaurus anatomicus: a morbid guide / edited by Joanna Ebenstein.

Description: Cambridge, Massachusetts : The MIT Press, [2022] | Includes bibliographical references and index.

Identifiers: LCCN 2020047135 | ISBN 9780262046039 (hardcover)

Subjects: LCSH: Human anatomy—Early works to 1800.

Classification: LCC QM21 .R982 2021 | DDC 611—dc23

LC record available at https://lccn.loc.gov/2020047135

10 9 8 7 6 5 4 3 2 1

Contents

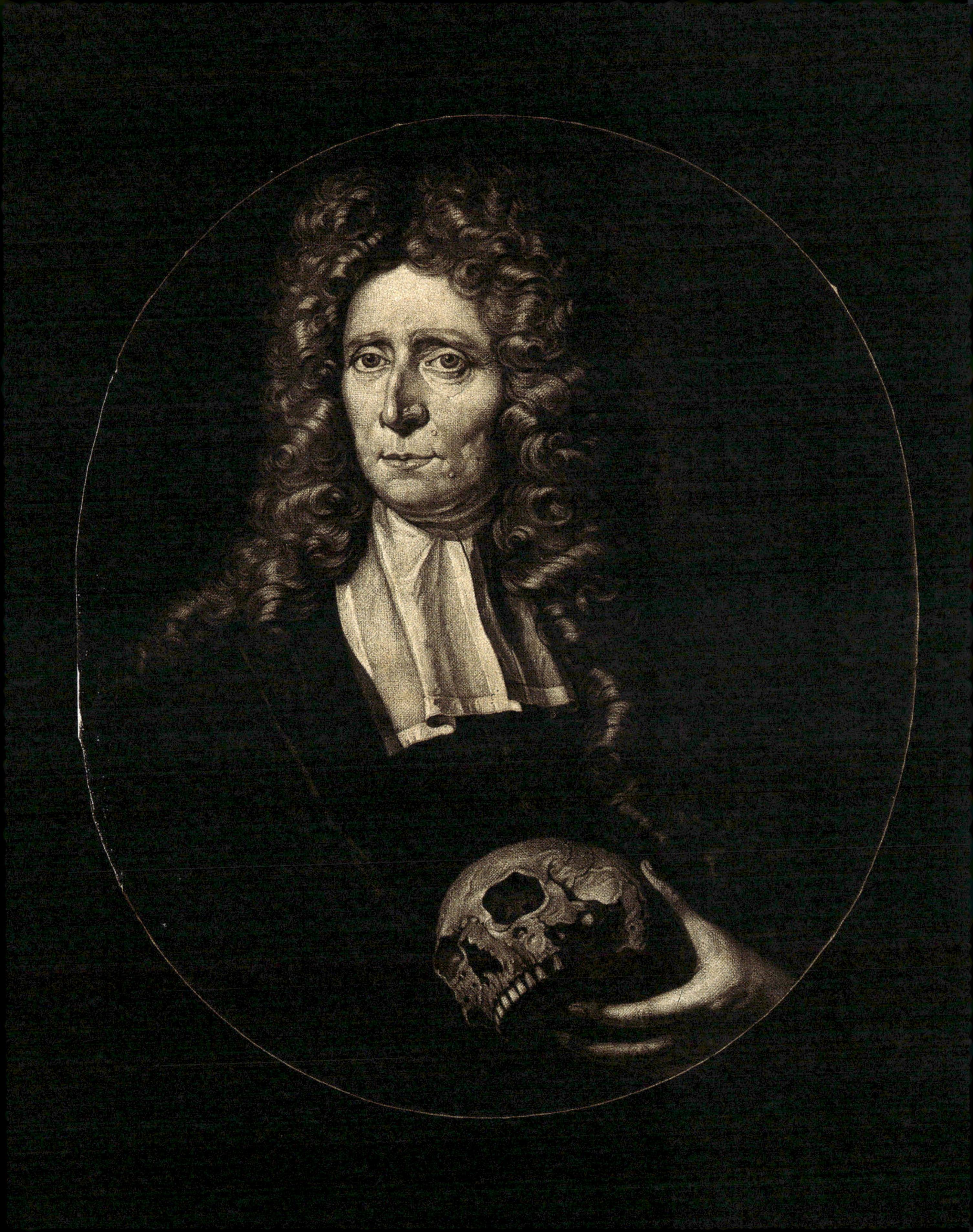

Frederik Ruysch Timeline

1638 (March 23 or 28): Frederik Ruysch born in The Hague, The Netherlands, to Hendrik Ruysch and Anna van Berchem

1654: Hendrik Ruysch dies; following this, Frederik Ruysch takes a job as an apothecary's apprentice in order to assist his mother and five siblings

1661: Frederik Ruysch opens apothecary shop in The Hague and marries Maria Post, daughter of the architect Pieter Post

1664: Receives medical degree from University of Leiden; begins medical practice in The Hague; Ruysch's daughter Rachel Ruysch born; she later assists her father with his preparations and becomes a famous still life painter

1666: Appointed as city anatomist of Amsterdam and praelector of the Surgeons' Guild; moves to Amsterdam soon after

1669: Cornelius Huijberts (Huyberts), illustrator of *Thesaurus anatomicus*, born

1672: Ruysch becomes city obstetrician and trainer of midwives; this provides him with access to miscarried embryos and fetuses, which he uses in his specimens

1679: Becomes doctor to the court of justice, which includes forensic investigations; this provides him access to the bodies of executed criminals, which he uses to create instructional anatomical specimens

1683: Jan van Neck, a Dutch golden age painter, completes his portrait *The Anatomy Lesson of Dr. Frederik Ruysch*

1685: Ruysch becomes professor of botany at the Amsterdam Athenaeum Illustre, which later becomes the University of Amsterdam; his role includes supervising the botanic garden

1691: Amsterdam's Surgeons' Guild opens a new anatomical theater, supervised by Ruysch, at the Waag (weighing house) on Nieuwmarkt square

1697: Peter the Great makes his first visit to Ruysch's cabinet

1701–1716: Frederik Ruysch's *Thesaurus anatomicus* published in ten volumes

1710: Frederik Ruysch's *Thesaurus animalium primus ... / Het eerste cabinet der dieren, van Frederick Ruysch* published

1716: Peter the Great visits Ruysch's cabinet for a second time

Frederik Ruysch, mezzotint by P. Schenck after Juriann Pool, 1665 or 1666-1745. From R. Burgess, Portraits of doctors & scientists in the Wellcome Institute, London, 1973. *Courtesy of The Wellcome Collection*

1712: Cornelius Huijberts (Huyberts) dies

1717: Ruysch sells his collection and his preservational formula to Peter the Great for 30,000 guilders

1720: Elected to the Royal Society of London

1721: Frederik Ruysch's *Opera omnia anatomico- medico- chirurgica* (Collected anatomical, medical, and surgical works)

1731 (February 22): Frederik Ruysch dies

1824: Italian poet, philosopher, and essayist Giacomo Leopardi publishes "Dialogue between Frederick Ruysch and His Mummies"

1831: French realist novelist and playwright Honoré de Balzac publishes his novel *La peau de chagrin*, in which the protagonist stumbles upon Ruysch specimens in a curiosity shop

Introduction

JOANNA EBENSTEIN

When we concentrate on a material object ... the very act of attention may lead to our involuntarily sinking into the history of that object.
—Vladimir Nabokov, *Transparent Things*, 1972

HISTORY IS A SECRET: A SIREN SONG OF A FOREIGN PAST THAT POSsesses us with a longing to *know.* The past is, however, ultimately unknowable, and always out of reach. All historical knowing is based on an act of human imagination. We can never *truly* understand the minds of those in other time periods, though certain compelling stories and objects nevertheless beckon us to embark on this unattainable quest.

I have spent a great deal of my professional life drawn in by the lure of history's secret worlds. One of the enduring subjects of my passionate fascination has been the work of Dutch anatomist, artist, and preparator Frederik Ruysch (1638–1731). Ruysch is an intriguing figure. His best-remembered works are still-life tableaux featuring fetal skeletons posed, as if alive, in frozen landscapes of preserved human remains.

These tableaux were intended to be, at once, objects of art, science, and metaphysics. Today they seem so strange that we can hardly believe they existed, much less that they were popular and beloved objects. How can we possibly understand the work of a man who was a brilliant scientist, technician, and physician, a religious moralizer, *and* an artist whose primary form of expression was the medium of human remains? And what does it say about our own era that his masterpieces, so admired only a few hundred years ago, now seem perverse, bizarre, and completely impossible? These are some of the questions that this book sets out to explore.

Ruysch exhibited his celebrated tableaux, along with hundreds of other anatomical specimens and sundry *naturalia*, in a cabinet he made available to the public in his Amsterdam home. Here they shared shelf space with his celebrated wet specimens: preserved human body parts in glass jars which were unique not only for the poetic bits of lace with which they were artfully augmented, but also for the illusion they gave of being still alive—rosy with the flush of life. Famously, Peter the Great of Russia was so enchanted with one of Ruysch's preparations that he kissed it, before going on to purchase the entire collection. The affective power of Ruysch's anatomical specimens was so great that they were sometimes referred to as "Rembrandts of anatomical preparation." They

Image on facing page: "Homo ex Homo," an illustration from Johann Jakob Scheuchzer's Kupfer-Bibel, in welcher die Physica sacra . . , *better known as* Physica sacra *(1731–1735). This series of books explores the physical world with a lens both scientific and religious, not unlike that of Scheuchzer's near-contemporary Ruysch. The fetal skeletons seen here are based directly on illustrations from Ruysch's books.*

continued to fascinate for centuries, making appearances in a novel by Balzac (*La peau de chagrin*, 1831) and inspiring a nineteenth-century literary fantasia about an encounter between the anatomist and his collection of dead children come back to life (Giacomo Leopardi's "Dialogue between Frederick Ruysch and His Mummies," 1827, reprinted in this volume).

From the moment I first saw Ruysch's work, it exerted an irresistible pull. I found myself wishing I could read his *Thesaurus anatomicus,* a heavily illustrated guide he published about his home museum in the early eighteenth century. It seemed to me that it would illuminate the almost unimaginably different approach to science, art, religion, museums, and collections that characterized his time, in a way that no academic article could ever fully capture. I had read that Ruysch juxtaposed scientific descriptions with lyrical responses to the objects penned by distinguished visitors. Memento mori mottoes followed rigorous documentation of each object's provenance. The subjective and the objective, rational and emotional, the religious and the scientific coexisted here effortlessly, for one brief historical moment, side by side.

Ruysch published in Latin and Dutch, and there were no English translations available. I developed a strong conviction that there should be one. I wanted to read it, and so would the community I had cultivated with my Morbid Anatomy blog, library, event series, and museum. It would also be an invaluable resource for historians of science, medicine, museums, and collections. For many years, I tried in vain to make this project happen. Then I met Matthew Browne of the MIT Press, whose publishing interest is the domain where science meets the arts and humanities. Soon after meeting him, I wrote a proposal for this book. Within days the project had been approved, and now, here in your hands, is the final product.

The centerpiece of this book, and its *raison d'être*, is the first English translation of much of Frederik Ruysch's *Thesaurus anatomicus*, complete with its full set of illustrations, drawn from the collections of the Bernard Becker Medical Library of Washington University at St. Louis's School of Medicine and London's Wellcome Library. The translator, Richard Faulk, has done a magnificent job of animating Ruysch and his world, finding the warmth and personal voice of this distinguished yet down-to-earth man. I have found reading it even more of a revelation than I had hoped.

As Ruysch's work has been rendered so strange by the passing of time, it seemed important to provide some context. To that end, the book begins with a collection of essays exploring many aspects of the man and his work by a range of international scholars, artists, and curators. Much of it reveals a passion that transcends the merely academic.

First, we have Luuc Kooijmans, author of Ruysch's definitive biography, who provides an overview of his life, his collection, and the main themes in his work. Next is an essay by medical artist Eleanor Crook, who brings to bear her

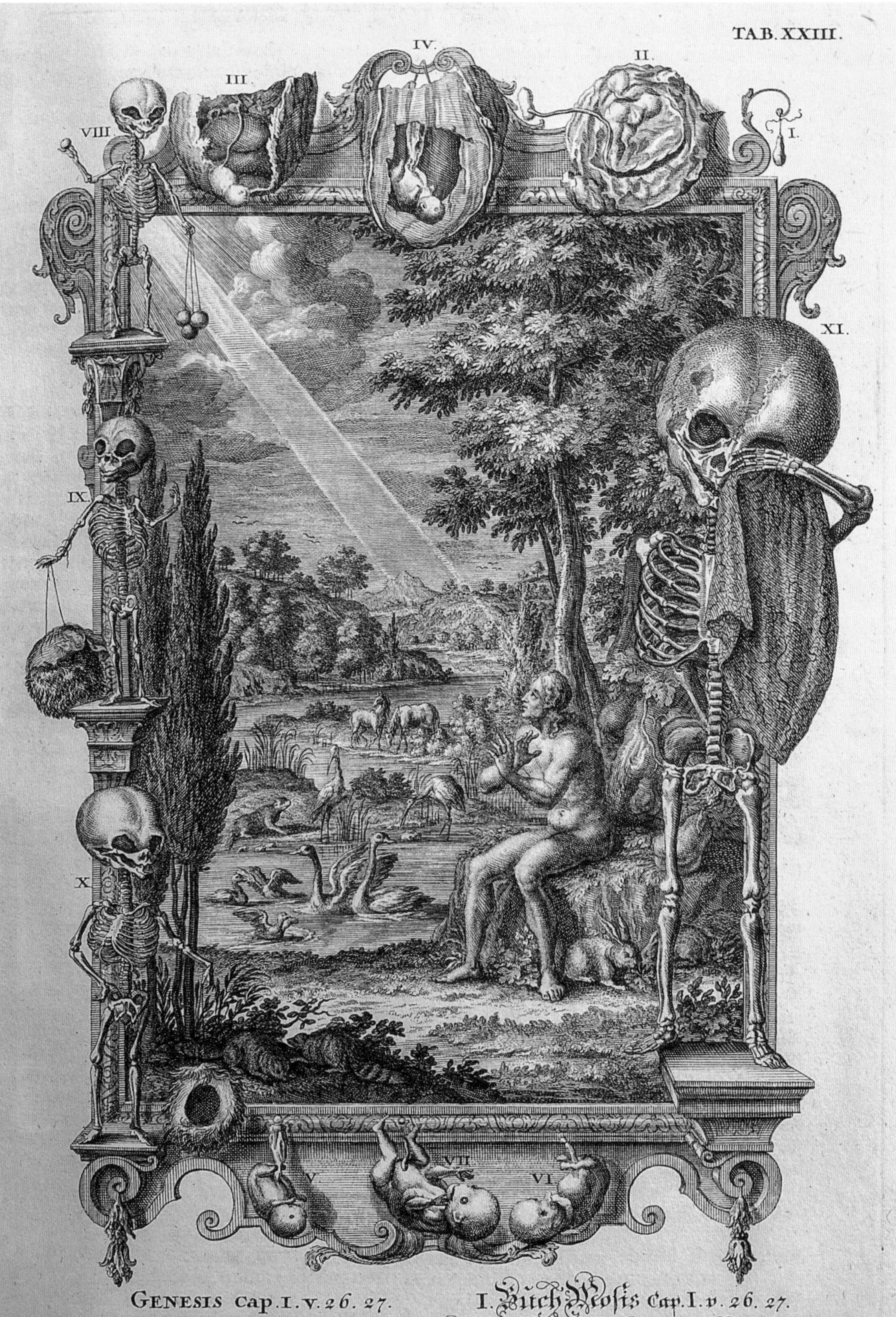

GENESIS Cap. I. v. 26. 27. I. Buch Mosis Cap. I. v. 26. 27.

Homo ex Humo. Erschaffung und Zeugung deß Menschen.

unique perspective as a maker well-versed in the subtleties of the dead human body and the depiction of its interior. Her piece is an ode to the paradoxical pleasures bestowed by this artist of death, with a focus on the virtuosity of his craft that engenders the illusion of life. Following this, we have cultural studies professor Bert van de Roemer, who writes of his fascination with the fluidity of the dialogue between art and science that one sees in Ruysch's work, and his own attempts to pictorially recreate Ruysch's museological cabinets. Curator Willem Mulder writes of his work conserving some of Ruysch's most famous specimens for the Kunstkamera in St. Petersburg, and the challenge of matching specimens with textual descriptions. Philippe Comar, one-time professor of morphology at Paris's École des Beaux-Arts, presents an astounding seventeenth-century fetal skeletal tableau from the École's collection, probably the closest we will get to seeing an original Ruysch in the flesh, so to speak. Laurens de Rooy, historian of medicine and director of Amsterdam's anatomical museum, the Museum Vrolik, locates Ruysch's work in the shifting ascendancies of rationality and romanticism in the natural sciences. And finally, philosophy professor Stephen Asma situates Ruych's work in the contexts of Enlightenment and esoteric thought.

I'm also delighted to include a translator's note by Richard Faulk, in which he lays bare the complicated nuances of the translation process. And, finally, we include the short piece referred to above, Italian writer Giacomo Leo-pardi's "Dialogue between Frederick Ruysch and His Mummies," originally published in 1827, in which Ruysch's specimens come back to life for a brief moment to discuss with him the mysteries of death. We are very fortunate to have specially comissioned brand-new illustrations which bring this piece to life, compliments of the multitalented artist (and essay contributor) Eleanor Crook.

Ruysch, it is clear, has acted as muse to a diverse group of thinkers and creators for centuries, despite the loss of his most famous tableaux. It is my hope that this book will extend his legacy, bringing his unique charms to a new audience ripe for anatomical enchantment.

While we can never truly know the past, this compendium brings it a little closer, allowing us to breathe, if but for a moment, the same air as this enigmatic "artist of death." It invites us to imaginatively enter a time—not so long ago and yet a world away—in which life and death, art and science, and religion and medicine had not yet been estranged: in which the approach to understanding life and its mysteries was not just to examine its disparate parts, but also to celebrate its connections, and in which we were encouraged to wield all aspects of our humanity in the service of understanding, and communicating, our discoverable world and all of its mysterious beauty.

Artist of Death

LUUC KOOIJMANS

OF DEAD BODIES HE HAD SEEN QUITE A LOT, AND NOT ALL OF THEM had come to their end in one piece, but what he saw now baffled him: a room in which the entire human body was displayed as if it were a collection of special stones or shells. And not only skulls and bones, but also hearts, brains, uteri, and countless other parts of the body that you never saw and that should have perished long ago or at least should stink unbearably. But they didn't smell at all. They looked like they could be put back into a body at any moment.

It was 1697, late in the summer. Peter the Great, the young tsar of the Russian Empire, was traveling through Western Europe, because he wanted to see with his own eyes what science and technology were available in the western world. When his hosts in Amsterdam suggested that he visit a doctor to see an anatomical collection, he was curious. But what he saw surpassed all his expectations. Surprised, he looked at a whole series of fetuses—some of them full-term, but also tiny ones, no bigger than a grain, which you had to look at with a magnifying glass. You could follow exactly how a human being developed before it was born.

He also saw a much bigger child, a child who had died after seven or eight years of life. But *did* it die? It had been placed in a tomb, beautifully dressed, with a lace collar and a garland of flowers around its head. It was as if it was still alive, as if it was asleep and could awaken at any moment. Peter was touched. He hugged and kissed the child.

Peter the Great may have expressed it in a remarkable way, but his admiration for Frederik Ruysch was by no means unique. A hundred years after his death, Ruysch was still famous as the artist whose preparations, instead of disgust and shock, aroused awe and tenderness.

At the beginning of Honoré de Balzac's novel *La peau de chagrin* (1831), the young protagonist is wandering around Paris, planning to put an end to his life, when he decides to go into a curiosity shop. There he encounters what Balzac describes as an "enchanting creature," the embalmed body of a child, which reminds him of his happy boyhood. This "sleeping" child turns out to be a remnant of the collection of Frederik Ruysch, who had been dead for exactly a century when Balzac wrote this novel.

The Italian author Giacomo Leopardi did not find the preparations horrifying either. In his "Dialogue between Frederick Ruysch and His Mummies" (1827), Ruysch is awakened at midnight by his cadavers, who have come to life in his studio and are singing in chorus. Ruysch, watching through a crack in the door, exclaims, "Who has been teaching these dead folks music, that they thus sing like cocks, at midnight? Verily I am in a cold sweat, and nearly as dead as themselves. I little thought when I preserved them from decay, that they would come to life again." He then enters his studio and says, "Children, what kind of game are you playing? Have you forgotten that you're dead? What's this racket all about? Has the tsar's visit gone to your head?" One of the dead tells Ruysch that they can speak for a quarter of an hour, so he asks them for a brief description of what they felt when they were at death's door. They assure him that dying is like falling asleep, like a dissolving of consciousness, and not at all painful. They declare that death, the fate of all living things, has brought them peace. For them, life is but a memory, and although they are not happy, at least they are free of old sorrows and fears.

Ruysch's main trick was heating white wax and injecting it into blood vessels in liquid form. Once it cooled and set, he would have a dissectible preparation. By staining the wax red, he managed to give bodies and organs a lifelike tint. The result was amazing. He used his preparations for teaching surgeons and midwives, but there was so much interest in them that he set up an exhibition. It was the first time that people could properly see human internal organs. The exhibition soon became a major attraction.

The museum was more than simply a collection of anatomical evidence. Those who entered were immediately confronted with a tomb containing various skeletons and skeletal remains. Among them was the skull of a newborn baby placed in a box, next to a sign with the motto: "no head, however strong, escapes cruel death." The tomb also contained the skeleton of a boy of three, holding the skeleton of a parrot, which had been placed there as an allusion to the saying "time flies."

Although the admonitory captions were very much in the established tradition of anatomical presentation, the museum was quite unique in that Ruysch had made an effort to give it an attractive design. Amidst the little skeletons in the tomb, for instance, was the embalmed body of a fetus of seven months. Its quite natural color already made the sight a little less unpleasant, but Ruysch had beautified the child in other ways as well, by putting a bouquet in its hand and a crown of flowers on its head. The flowers, too, had been preserved so that they would keep their petals and their bright color.

Visitors were confronted with the skeletons of a child of four with a toy in its hands, a five-year-old holding a silk thread with an embalmed heart dangling from it, and a girl drying her eyes with a pocket handkerchief.

Decorations, memento mori images, and *vanitas* symbols put the horror of death in perspective by stressing the transience of life, by showing that the body was no more than an earthly frame for the soul. After death it no longer served its purpose; only an anatomist could make it useful to the living.

Of course, the real goal of the anatomist was not to amaze his audience, even though that ambition, too, could be justified, particularly by arguing that it would impress the viewer with the wondrousness of God's creation. The ultimate objective of anatomy was to increase man's knowledge of the structure and workings of the human body. Ruysch had developed his skills to be able to make structures visible that would otherwise have remained invisible.

Toward the end of the seventeenth century, after thirty years of practice, Ruysch, assisted by his son, managed to perfect his preparation method, which, as he said, now made "the tiniest parts of the human body clear to the eye." Key to the process was the injection of a substance that would not congeal until it had penetrated the tiniest of blood vessels. Liquid wax went a fair way toward this objective, but not as far as Ruysch wanted; he had constantly been looking for a better substitute. Once he had solved that problem, he began to make many new anatomical discoveries, and he developed the ability to create preparations that differed little in appearance from living tissue. Most of them now were stored in glass jars and bottles, in a remarkably clear liquid he called *liquor balsamicus*, which preserved their lifelike color and elasticity. Whereas they used to become hard and stiff, with faded color, now they were kept bright and supple.

When Ruysch first made his results public, his technique was considered akin to sorcery. Some people simply refused to believe their eyes. Ruysch was accused of using trickery to make his preparations more attractive. He was often criticized for the way he presented his anatomical material. What was the point of all that embellishment, he was asked, and he countered by demanding to know why people spent so much money burying bodies that were already worm-eaten. His reasons were clear: "First of all, I do it to allay the distaste of people who are naturally inclined to be dismayed by the sight of corpses," he said, but (and this was why he didn't want to be accused of cheap tricks) he also saw a clear connection between the appearance of his preparations and their scientific validity, for he claimed the ability to restore a body to the state it was in before death. That his corpses seemed to be asleep was not just amazing, it was also significant.

Because his new technique enabled him to work more efficiently and to make better and more beautiful preparations, he decided to reorganize his museum. In doing so, he concentrated more than ever on the presentation. The collection was placed in a number of cabinets that filled three rooms. As before, the arrangement was not systematic: Ruysch turned every cabinet—which he called a *thesaurus*, a storehouse of knowledge—into an individual work of art, consisting of different kinds of preparations in unique combinations. The centerpiece of every cabinet was an anatomical still life placed on a bed of bladder-, kidney-, and gallstones, from the midst of which rose "trees" of dried blood vessels filled with a red waxlike substance. Among these stood tiny fetal skeletons. They still delivered their grave message to the visitor, but by now they did it with a sense of humor.

Such a combination of seriousness and humor fit in with a long-standing tradition. Sixteenth-century publications abounded in illustrations of living skeletons stepping into tombs, or using bones as drumsticks. In a way, Ruysch's compositions were subtle, three-dimensional versions of those anatomical illustrations showing skeletons in dramatic poses placed in curious settings. Numerous publications on anatomy—the famous atlas of Vesalius, for example—contained pictures of skeletons portrayed as gravediggers, or hanged criminals. But to present skeletons as characters in a *tableau non-vivant* was highly unusual, and an indication of the extent to which Ruysch had distanced himself from his material. In one of his compositions a skeleton says: "even after death I'm still attractive!"

Although the collection reflected his search for answers to scientific questions, and could be used to answer such questions—at least when Ruysch could find the preparation he was looking for—it was largely an end in itself. Ruysch did not order his work according to specifically formulated problems. Instead, he confined himself to describing his collection and recording the observations his injections had enabled him to make.

He maintained that he was merely gratifying the desire to observe the miracles of God Almighty, but in fact his motivation was twofold: not only did he exalt the human anatomy as a wondrous product of creation, but he presented himself as a veritable artist of death. His display of his anatomical virtuosity contained the veiled message that he—and he alone—was able to defy death to the extent that he could make a corpse look like a living body. He always emphasized the natural form, color, and flexibility of his prepared bodies, which differed from live ones only in their lack of movement. He was convinced that his art had given him knowledge—and therefore a special status—that others would never be able to attain. This explains his reluctance to divulge his working methods, for it was only by maintaining strict secrecy that he could remain the sole intermediary between the living and the dead.

The secret of his methods has never been unraveled. But some of his preparations still remain, in Peter the Great's Kunstkamera in St. Petersburg. And it is still the soft pink cheeks and the three-century-old baby fluff on children's heads that attract the most attention. The scientific significance of his preparations may now be limited to history; he has continued to live as a death artist.

The Pervasive Spirits of Frederik Ruysch: A Sculptor's Appreciation

ELEANOR CROOK

CONSIDERED AS A SCULPTOR, FREDERIK RUYSCH COMBINED AN UNcanny flair for material handling with a flamboyant expressiveness in symbolism. Concealing his hard-won preservative techniques behind an illusion of realism and scientific transparency, he broadened the vocabulary of the anatomical macabre while exploring some of the body's most narrow and tortuous tunnels.

In offering my reflections, as a sculptor in wax with a special interest in anatomy and pathology and a long-time admirer of Ruysch, I should explain that I take certain liberties of thought. An artist can claim the freedom not to be bound by the rules of historical writing or the original religious and moral context of the specimens and presentations; can take the objects on their own terms, considering how they may be read in our moment, as survivors of a past we can no longer enter, messages from a period that helped shape our own culture but would not recognize our predilections and preoccupations. An artist can consider such objects possessively, subjectively, as sources of inspiration, suggestions for plucking the strings of human emotion, entry points to the uncanny valley. So: first, the Ruysch jars, whose contents I have mimicked with wax silk and liquids in an attempt to capture something of the surreal charm of the infants, body parts, and treasures he stored for us under an ancient meniscus.

The antiquity of the Ruysch wet preparations has its own power. The surviving bodies and parts, mummified or drowned in their *liquor balsamicus* for three centuries and turning lazily in their post-mortem slumber as we lift the jars to inspect them with a mixture of wonder and dismay, have become travelers from an antique land, physically outlasting their contemporaries and descendants, outlasting for all we know the expectations of their preparator, outlasting ourselves. I appreciate the Ruysch specimens for the atmosphere and ideas they convey to us now, and my historical angle is one of technical interest, material speculation: what understanding of bodily tissues drove his accomplishment in arresting decay, what knowledge of exotic and of familiar chemical substances allowed him to suspend animation so successfully.

The clue is in an early treatise by Ruysch, *Dilucidatio valvularum in vasis lymphaticis et lacteis* (The Hague: H. Gael, 1665). Ruysch was not the discoverer of the lymphatic system, but he studied and dissected it minutely. An almost invisible, intricate and fine circulatory system for fluids, immune cells, and waste

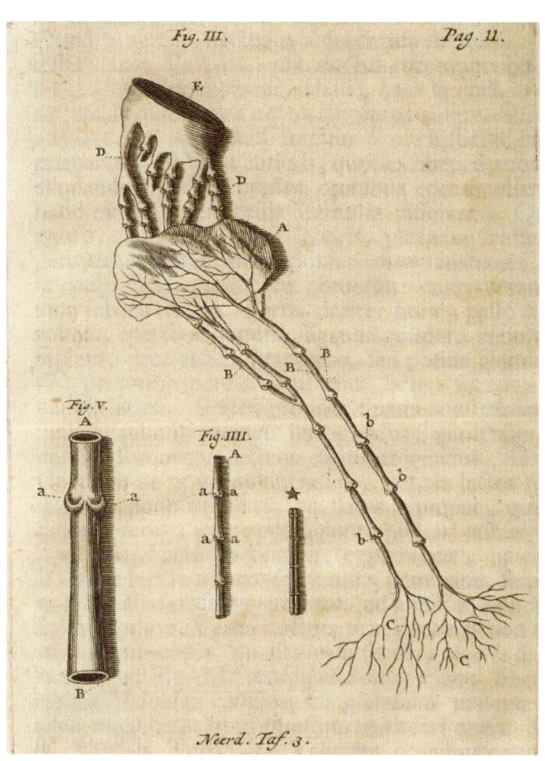

Image right: Lymph vessels and lymph node from the intestinal mesentery of a horse, prepared by Frederik Ruysch, from his Dilucidatio valvularum in vasis lymphaticis, et lacteis ... Accesserunt quaedam observationes anatomicae rariores, *1744*

products, it transports a clear liquid to and from the blood vessels, and you will only notice yours if it fails—yet it pervades all of you. Ruysch discovered its system of flow to be based on valves. His book, with its engravings of the lymph nodes, valves, and tubules in animal organs, may not be the most visually arresting of Ruysch's productions, but the research within it made possible his entire output of wet, dry, and injected preparations.

The secret is an astonished understanding of how flesh is structured, that what appears to be a generally soft solid substance is in fact a spongy, highly infiltrated world of tubules, of osmotic flow and ebb, seeping and boiling with liquid and gaseous exchange, a living forest floor of vanishingly small vessels, capillaries, interconnecting tunnels and loops, a constant chemical seethe, under and especially within our skin. There is no corner of an animal's tissue that the liquid does not reach and enliven, and as anyone who has been bitten by a Malayan pit viper will tell you—as the blood clots in his veins like cooling wax—we underestimate the speed with which molecules whirl through our systems. The understanding is one of liquid behavior, liquid motion, and it is no coincidence that a Dutch anatomist, familiar with that country's marshy lowlands and those navigable cities with their arterial canal systems, sluices, and locks, should have had this insight. Ruysch had identified lymphatic structures through the painstaking injection, under a microscope, of the tiny vessels with a waxy substance, and even more fantastically by tying the vessels with thread and injecting air. This research required superhuman patience and delicacy of handling. He himself stated that some of the lymphatic vessels are as fine as cobweb, and they are as easily destroyed by even light touch.

Wax injection of blood vessels as a preservative and sculptural procedure had a noble precedent in the anatomical injections of the brain by Leonardo da Vinci, was pursued in the later seventeenth century by Jan Swammerdam and Ruysch himself, and was developed subsequently for dissection and for corrosion casting of the tissues by acids. It was imitated by the use of wax modeling by syringe in later anatomical waxes all over Europe but perhaps perfected by the Florentine wax studios of the Museo La Specola in the late eighteenth to mid-nineteenth century. It sounds simple enough, but in practice requires great control of temperature, swift handling of rapidly deteriorating morbid materials, fine instruments, pure waxes, oils, fats, and pigments, and the sensibility of a jeweler. I emphasize it because when I see a rosy-cheeked Ruysch infant pleasantly dozing in his silk bonnet in a jar, I know from my experience with materials that what I am really seeing is a miracle of controlled perfusion and injection—perfectly selected materials, a deft hand controlling the plunger on a superbly engineered seventeenth-century brass syringe with a needle whose bore must have been hair-thin and utterly uniform; a child which must have been warmed in fluid to the precise same temperature as the molten wax fat and pigment mixture to prevent it coagulating halfway into the vessels, whose human fluids must have been drained off and replaced by the mixture completely, and then the little body steeped in highly distilled alcohol mixed with some mysterious ingredients involving a repertoire of pepper, clove, tannic acid, and who knows what obscure magic thing. The wax mixture fills the blood vessels, hence the blush, but the alcohol fills the more minute spaces, the lymphatics and intercellular space. It is an illusion born of much knowledge, practice, and patience. The maker in me is happy to see it, and amused by the bravado of the technique, which hides the extremity of the intervention of the artist's hand by making it appear that nothing has been done to the specimen, that it is just as it was.

Yet even as my eye accepts that it is looking at a living face (see image on page 88) flushed with pink living blood (actually mercury sulphate, the cinnabar of the Florentine wax arteries) looking back through the glass wall with living, not clouded, eyes, a face which even carries a living facial expression (albeit in this case the expression of one who has awakened to a new endless existence, tentative, puzzled, unquiet), the disembodiment undercuts the sense of it living. I am forced to hold two contrary impressions: undeniably dead, undeniably conscious. Undeniably horrible, undeniably beautiful. Tender pathos, heartless scientific objectification. These competing experiences raise the tension of the encounter: depending on my sensibility it will end in delighted curiosity or appalled disgust, perhaps a heady cocktail of both, seasoned by technical admiration. Contemplating a Ruysch baby is certainly a guilty pleasure.

Then again the child—often an unborn, or stillborn—is in its element, a liquid capsule; yet this liquid, inimical to even the life of bacteria or fungus, charged with toxins, tannic acid, camphor, alcohols, changeless as a diamond, could not be further from the protective, nutritive amniotic fluid it replaces. I

Image on facing page: Two preparations of Surinam toads (Pipa pipa) *by Frederick Ruysch, presented in jars with lids decorated with shells, corals, and small amphibians. From his book* Thesaurus animalium primus *(Amsterdam: Joannem Wolters, 1710).*

become aware of the distortion, the refraction of the light that makes the pinkish cheeks loom larger than life against the curve of the vessel, that magnifies the specimen until it leers in my regard. It does not breathe as I breathe the air, yet where the liquor has evaporated, perhaps part of it does protrude into our world, our air, placed even closer to the threshold of the living. Then there is the decoration: lace, pretty braids, silks and tassels, baby bonnets, fancy foil around the jar lid, all of it (some supplied by Ruysch's daughter, the still life painter Rachel Ruysch [1664–1750]) wheedling to charm me, to distract me from the facts of death and amputation, to soften the blow of the anatomist's cold steel, to comfort me. Maybe this sweetening succeeded in its day, but to my present sensibility these frills have an equal and opposite effect, flirting coquettishly with my aesthetic sense and mawkishly accentuating the pathetic morbid spectacle. The Ruysch babies seem somehow humiliated by them.

The still life arrangements of coral, fish, shells, and certain small creatures, which used to top off some of the jars, now lost through fragility, we know from the engravings. These are the missing link between the liquid world and our own: they reach out and up into our air, preserved and arrested in form but not permeated by fluid, intricate and organic in our physical space, not that of the jar. It took me a while to understand the formal function they serve in the collections: more than decoration, they offer an alternative to the floating suspended animation of liquid preservation. A dry specimen can be posed like a puppet. Here, form is made crisp and permanent by careful drying, varnishing, and gluing, a different permanence from perfusion: color and hydrated shape are sacrificed for accessibility and an illusion of arrested movement.

The toad jars (image on facing page) contrast the submerged specimen with an action tableau atop each—toadlets leap in alarm away from a predator, spiraling around a turret of shells and corals. This is pure storytelling, narrative taxidermy such as would later become fashionable, a whimsical language of display and a rare example of kitsch in the Dutch golden age. One can believe that Rachel Ruysch had a hand in this assemblage, and in the more ambitious *vanitas* centerpieces of the cabinets which we know only through tantalizing engravings. As I have no way of knowing the extent of her contribution, I will refer to these assemblages as "Ruysch tableaux," allowing for the collaboration.

The wit and charm of the major *tableaux non-vivants*, as well as the elegance of their composition, have me entirely under their spell. Aside from the animation of the fetal skeleton poses and their inescapably mischievous grins, there is much to enjoy in the clever repurposing of calculi, vascular casts, penis linings, mesenteries, and membranes to present a harmonious and visually consistent landscape. The material aptness makes a sculptor's fingers twitch with haptic pleasure; the conceptual neatness of medium and message, form and content, is deeply satisfying and carries a moral weight. As has been observed

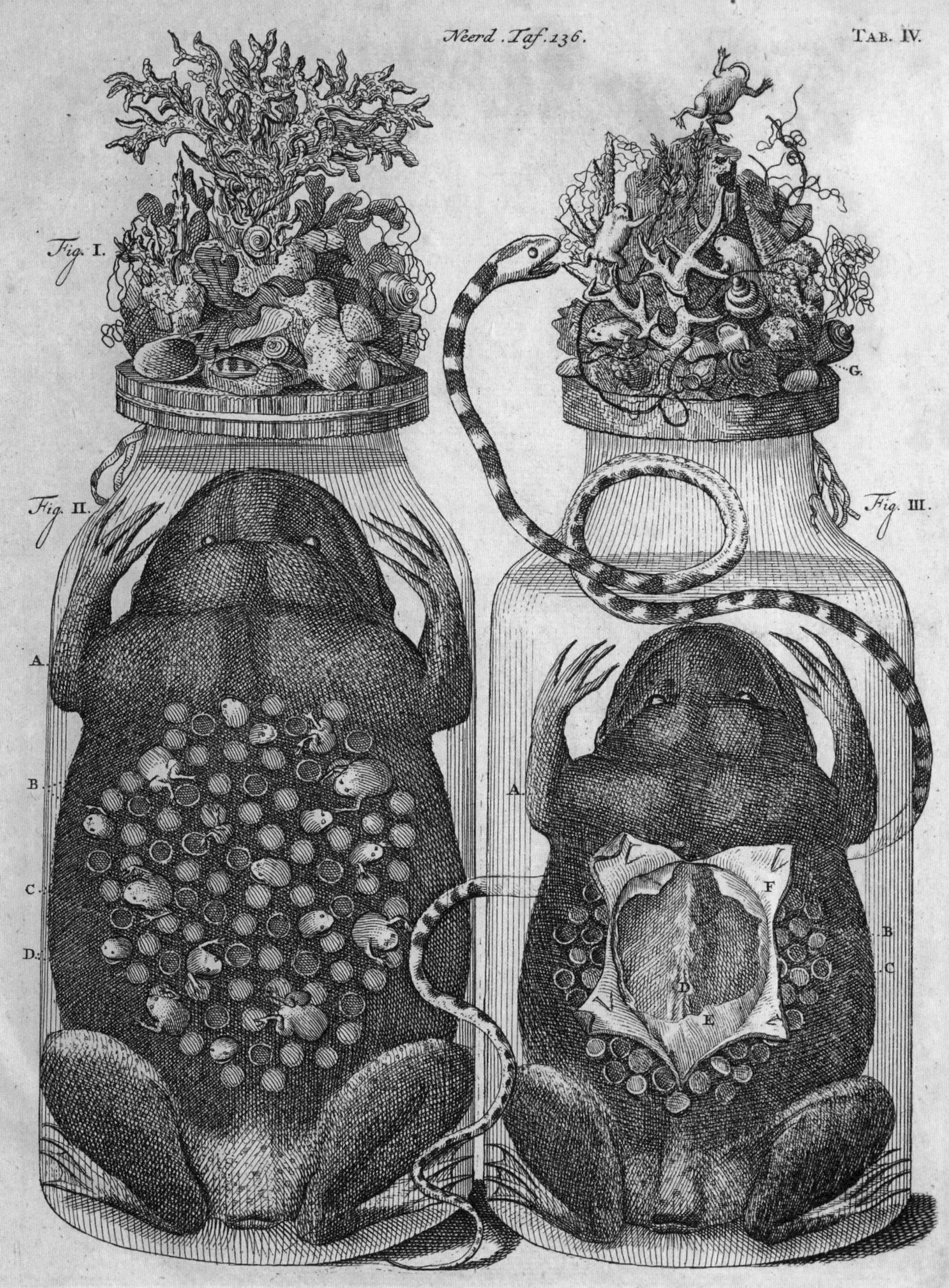

Neerd. Taf. 136.
TAB. IV.
Fig. I.
Fig. II.
Fig. III.
A.
B.
C.
D.
A.
B
C
D
E
F
G.

elsewhere, the use of human material as a sculptural medium is an ethically questionable one (even more so now than then) and an area where an aesthetic error can tip into a moral one, but the Ruysch tableaux walk that tightrope deftly, dignified by excellence of execution and composition, complexity of conception and … chutzpah.

Of course these are three-dimensional descendants of the animate skeletons of Vesalius and his predecessors in anatomical illustration, but it is the full incorporation of specimens and fragments into landscape that seems new; the body is its own world, its tissues can supply any tissue, it has an inexhaustible supply of forms, it is the microcosm of the universe that the Renaissance suspected. Handling and knowing the material so intimately has given Ruysch the visual ideas: the project of the tableaux is a fine example of hand-making as invention. The matter is the subject, the subject the matter, and both are transitory and doomed. Somehow this message is conveyed with a lightness and tenderness new and yet to be equaled in anatomical display; it is a great pity that these magical microcosms have not survived. There is nothing under any Victorian dome to compare with them, and even the modelers of wax tableaux do not present this human comedy with such inventive directness—that an unborn human should mourn for the dead poet! That a stillborn child should play Death's fiddle with an arterial bow and lament its fate! It is enough to make a person reach for a scrap of meningeal tissue to wipe away the *liquor balsamicus* of a tear. Be we sculptors, anatomists, historians, connoisseurs, mortal spirits all, these messengers from the laboratory of a magician in matter speak eloquently down the years of the wonder and pathos of our brief lives.

The Remainders of a Reconstruction: Visualizing Ruysch's Collection

BERT VAN DE ROEMER

FREDERIK RUYSCH WAS NOT ONLY A REMARKABLE MAN IN HIS OWN age; he also holds a remarkable position within historiography. His historical persona has been interpreted in divergent ways. Should we see him as an artist or a scientist? Was his main aim discovering and disseminating new anatomical knowledge, or was he more interested in making an impression with his spectacular collection? Was he an exponent of the baroque age, or an enlightened radical spirit?

In my research into Dutch collections of curiosities I am always intrigued by the gray areas between different categorizations and classifications. This applies as well to modern categories projected onto the historical persona as it does to the classifications used by those collectors to order their objects. In Ruysch's case, I am mainly interested in the vague area between "the artist" Ruysch, making enchanting and engaging objects, and "the scientist" Ruysch, searching for hidden knowledge about the anatomy of the human body. These two activities might seem at odds with each other by the light of current disciplinary divisions, but the intriguing thing is that Ruysch and his contemporaries did not feel any friction here. The idea that knowledge should be presented in a neutral and transparent, nonaestheticized way if it is to be considered "scientific" is relatively new and was not current in those days. Moreover, contemporary manuals encouraged collectors to make a combination of impressive and aesthetically pleasing presentations that at the same time offered an instructive, knowledgeable sight to inform the beholder. A good collection was ordered according to sound systematic-philosophical insights, but also with "splendor," which usually meant a geometrically well-balanced and neat whole. Only then would the assembled objects make a useful impression in the eyes of the beholder.

And Ruysch knew how to do this like no other. Had someone asked Frederik Ruysch whether he was an artist or a scientist, he would undoubtedly have raised his eyebrows, wondering what the questioner was talking about. It is a commonplace to state that "art" and "science" in the days of Ruysch were something different than they are today, but it is another thing to really try to put your finger on this difference. How, then, did these concepts function in the mind of Ruysch? Or rather: What did he mean when he used the words *konst* (art) and *kennis* (knowledge), and how does this relate to our modern concepts?

I will give two examples of how these concepts functioned differently. By many, Nature was seen as God's creation, a majestic artwork, something well thought out and contrived, and not as something that came about through a combination of natural laws and genetic mutations. Like an artwork, it was something constructed, and someone or something was responsible for that construction. This preconception would encompass a totally different attitude toward nature. On the other hand, Ruysch used the word *konst* many times in his writings, but almost always when he referred to his embalming technique, and not so much for the decorations he made. In this way, the word "art" was used to designate a technique or an operation that enabled man to supplement or improve nature; in the case of Ruysch, stopping the process of decay. My fascination with Ruysch comes from the fact that he—like other collectors from his time—felt no tension between his artistic and scientific practices and used the concepts differently and more fluidly. It is this fluidity that interests me and that I want to understand better.

How could I investigate the relation between the artful and the scientific in the collection of Ruysch? For a good comprehension of the activities of this collector, I decided, it was necessary to get a visual impression of the order of the whole collection as it would have appeared in the late seventeenth or early eighteenth century. Many of Ruysch's preparations survive in St. Petersburg, as he sold his collection to Peter the Great in 1717, but of course they are not preserved in their original state or order. Furthermore, there are no images of the rooms that held the collection that can give us an impression. There only exists a highly idealized representation of a room with cupboards in a frontispiece of one of Ruysch's catalogues (page 28, bottom image). So to get a view of how the embellished and artistic preparations were distributed over the whole collection, and how they related to the preparations that merely conveyed anatomical fact, I had to construct myself an image of what it might have looked like. I decided to start with the first catalogue published in 1691 in which Ruysch described the contents of his collection for the first time. The collection consisted in those days of eight cupboards with anatomical preparations and two cupboards with a herbarium. As Ruysch described every object in a rather detailed manner, cupboard by cupboard, shelf by shelf, it was quite easy to get the general picture in words, but what about a visual impression?

To accomplish this, I bought ten large cardboard sheets, one for each cupboard, and divided them horizontally according to the arrangement of the shelves. Subsequently, I took small index cards and sketched the objects diagrammatically as described by Ruysch. I pasted the cards on the cardboard sheets according to the order of the descriptions in the catalogue. Of course this way of working was hazardous, because you visualize things for which you only have textual information.

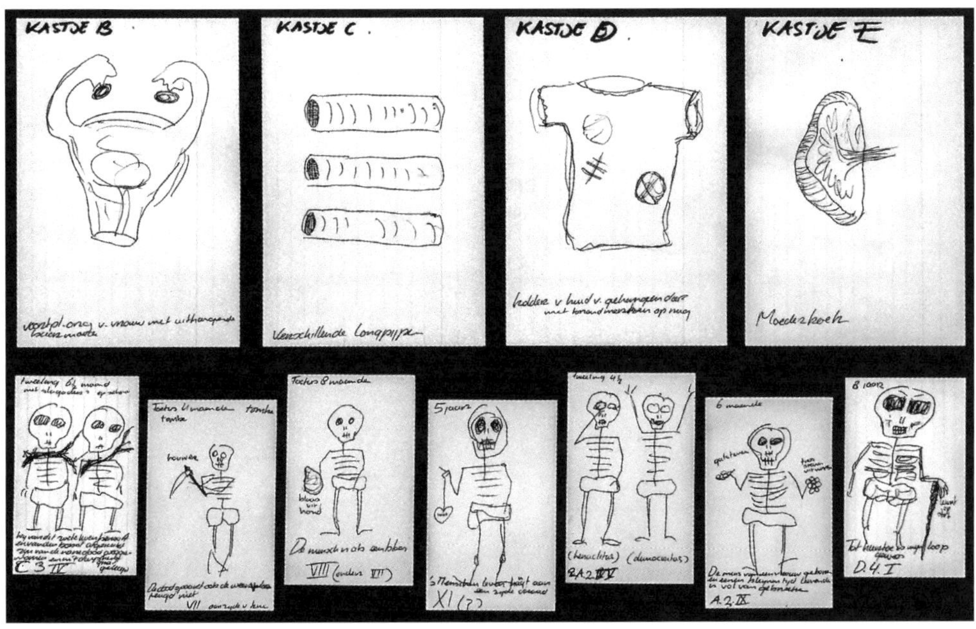

Top image: First frame with the remainders of my reconstruction of the collection of Frederik Ruysch.

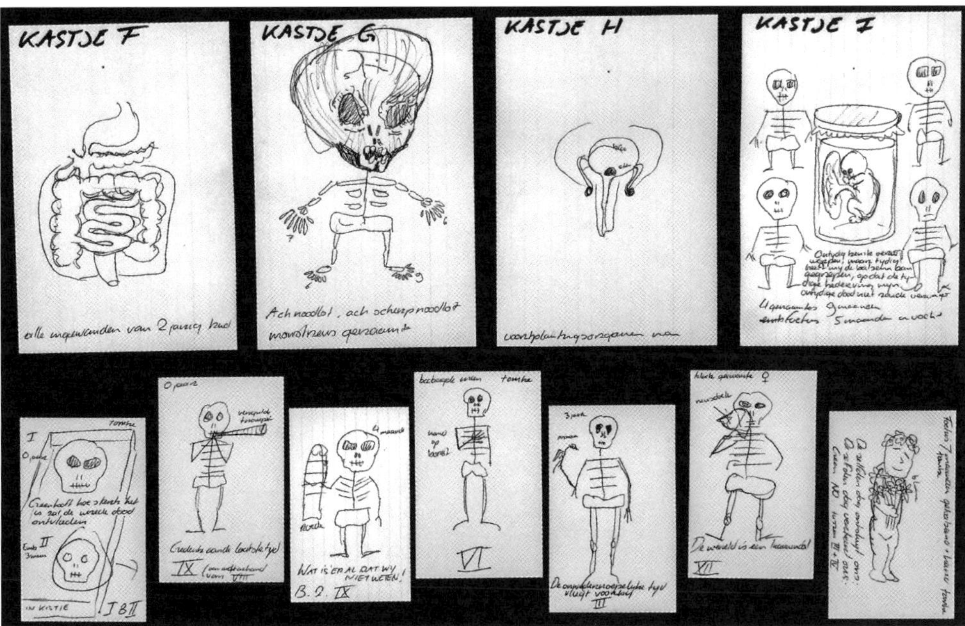

Bottom image: Second frame with the remainders of my reconstruction of the collection of Frederik Ruysch.

Sometimes I was more certain, because Ruysch had many preparations depicted in his earlier publication of anatomical remarks, the *Observationum anatomico-chirurgicarum centuria*. However, most were not visualized. But, as my aim was to get a general view of the distribution of the embellished preparations among the entire collection, and not exactly what the objects themselves looked like, I considered this way of working legitimate. At that time—

the first year of my PhD research in 1996—I had a very small room in the Institute of Art History on one of the Amsterdam canals. I attached the sheets to the walls and found myself surrounded by the collection of Ruysch. Regrettably, the cardboard sheets did not stand the test of time when my partner and I moved houses, but luckily over time some of the index cards had fallen loose from the sheets and scattered over the attic floor. My partner secretly collected and framed these remainders and created a special and very personal gift for my birthday (images on previous page). They now have an honored place in our corridor, and it is always a pleasure to look at them, remembering my first steps in my research into the conceptual framework of Dutch cabinets of curiosities.

The two frames each contain two rows of items. The upper rows with the larger cards show the contents of eight smaller cupboards, each containing one special display, which Ruysch had placed above the larger cupboards. The lower rows consist mainly of objects kept in the larger cupboards, or that formed part of a larger installation, which I will discuss later.

This first visual reconstruction led to some conclusions. First, it made me more aware of the different phases the collection went through in the long period of its existence. Ruysch lived a very long life and he prepared and collected for some sixty-six years. Often historical collections are taken for granted as more or less fixed entities, but in reality they changed continuously, as collectors were constantly acquiring, adjusting, and rearranging. In the case of Ruysch, the collection developed unceasingly through the years and was on two occasions revised considerably. In the last decade of the seventeenth century, Ruysch experimented with a new technique of wet preparation by injecting his *anatomica* with a secret mixture containing red-colored cinnabar, and hanging them in phials with thinned alcohol. This breakthrough led to a total reorganization of the collection in the last years of the seventeenth century. Another transition point was the sale of his collection to Peter the Great in 1717. After this, Ruysch started a totally new collection that grew rapidly.

Another, related thing I became more aware of while looking at my visual reconstruction was that in the first, relatively unknown phase, before 1700, the collection was comprised mainly of dry embalmed preparations placed separately on shelves. Of the 408 preparations mentioned in the 1691 catalogue only seven were described as "wet," meaning preparations preserved in alcohol. The last drawing in the upper row of the second frame, labeled "Kastje I" (little cupboard I), shows one of Ruysch's early attempts at his new technique.

It contained a fetus of about five months' gestation which, according to Ruysch, was preserved so well "in its liquid and natural state" that it seemed to live. He was famous for bringing dead material back to life, at least so it seemed. The fetus in alcohol was escorted by four "dry" preparations of fetal skeletons of nine months. He provided engaging objects like these with proverbial inscriptions, as if they were speaking to the beholder. This one said:

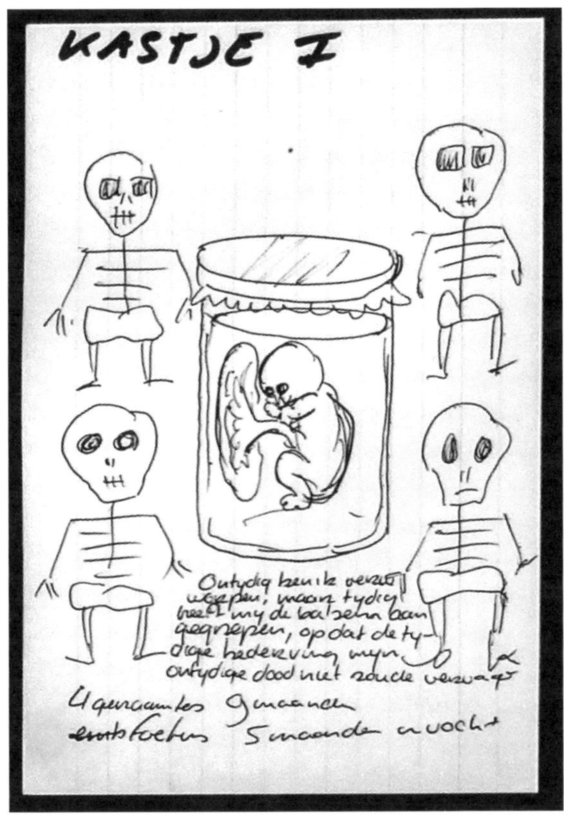

I was expelled prematurely, balsam received me in time,
So the premature death could not strike me with timely decay.

This was the last object described by Ruysch in the catalogue of 1691, and can be regarded as a prelude of what was to come in the new phase after 1700.

The visual reconstruction also showed that the amount of "embellished" preparations was relatively small. About one-fifth of the collection could be considered "artistic" in the sense that the items had some additional meaning bestowed upon them by adding attributes or inscriptions. These decorations consisted mainly of skeletons on wooden pedestals, holding attributes that expressed the vanity of life. They were scattered over the shelves and placed randomly between the dry embalmed preparations that merely showed anatomical information (see next page, top image, for an associative visual impression of this phase). The famous rocky landscape tableaux of stones, vessels, and skeletons that are reproduced almost obligatorily in studies on Ruysch were

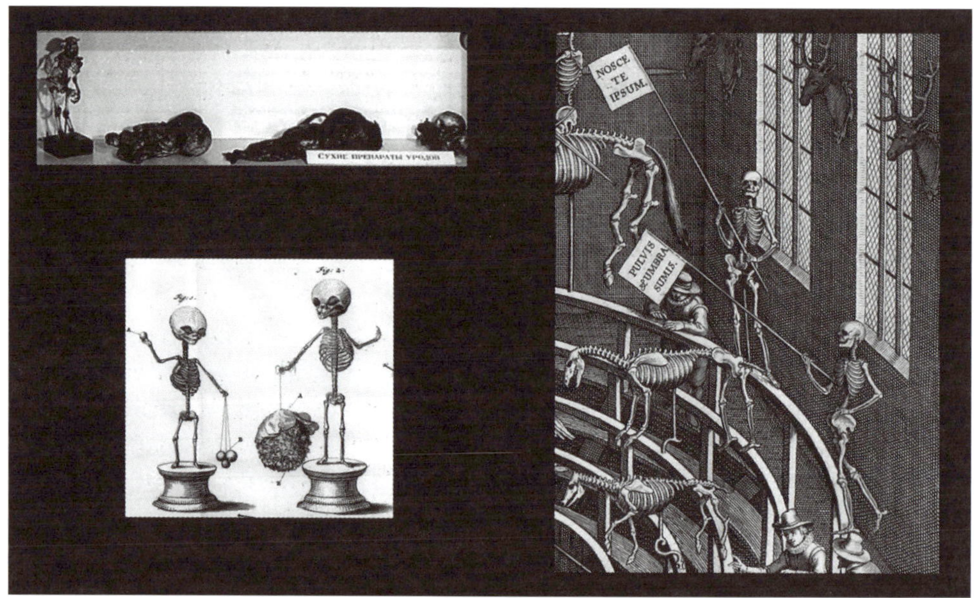

Top image: A visual impression of the first phase of Ruysch's collection around 1690. The shelves were filled with dry embalmed preparations, as shown in the upper left image, which were interspersed with fetal skeletons on wooden bases (image below left, from Ruysch's Thesaurus*). In a part of the room was a larger construction with several skeletons reminiscent of the decorations of the Theatrum Anatomicum (anatomical theater) in Leiden, where Ruysch had studied medicine (image on the right).*

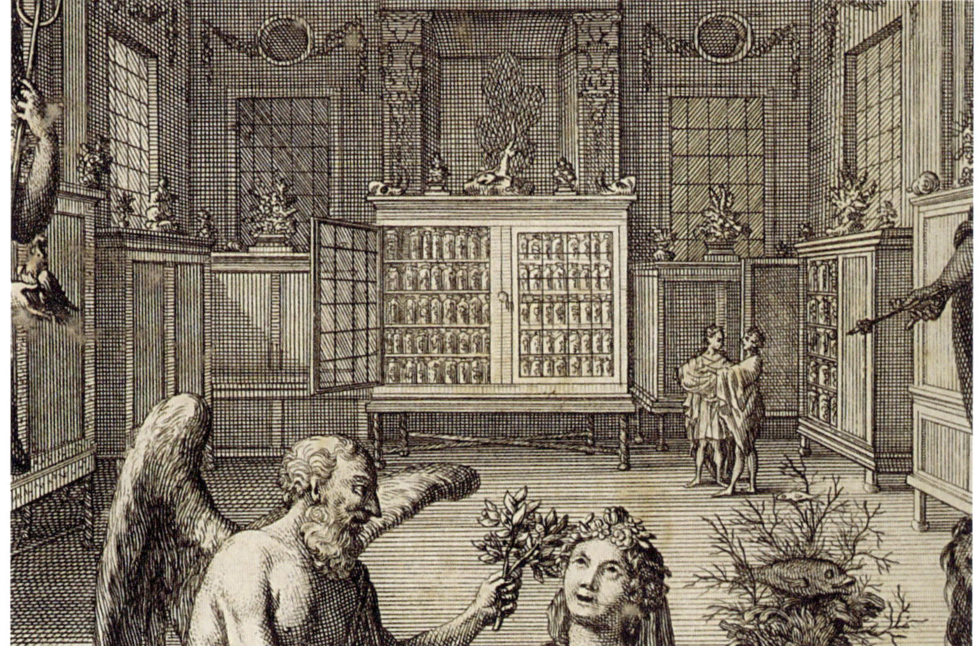

Bottom image: A visual impression of the second phase of Ruysch's collection around 1710. This is a detail of the frontispiece of the Thesaurus animalium primus *from 1710. The room is fictitious, but the ordering in the cupboard is most likely accurate.*

not present in this phase (see pages 115 and 148 for examples). After the reorganization around 1700, the disarray of embalmed preparations on shelves made way for a more harmonized and orderly preparation of neat rows of phials filled with wet preparations and thinned alcohol. An impression of this we can see on the print from the frontispiece of a later catalogue from 1710 (see the lower image on this page).

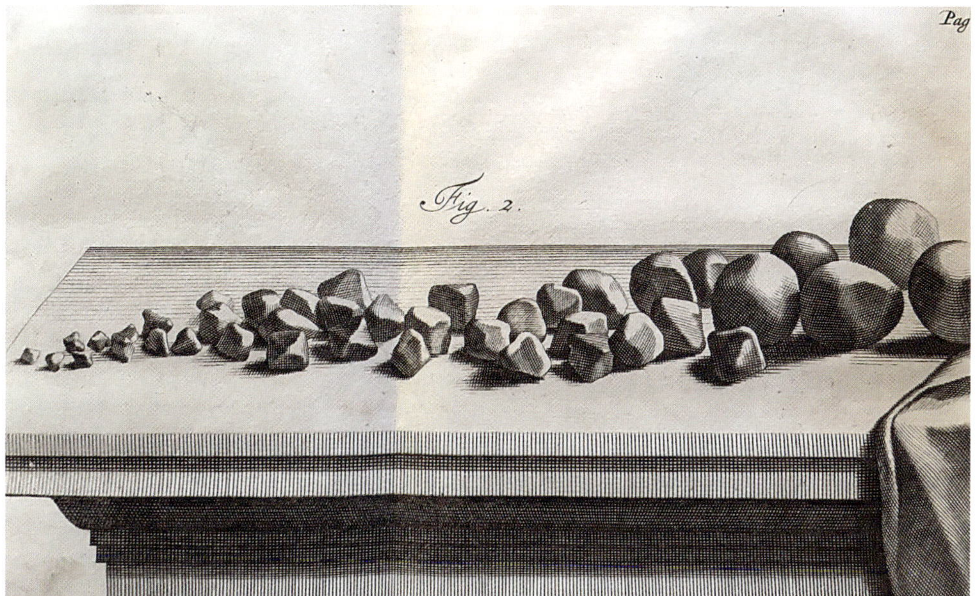

Image left: Forty-one stones cut from the bladder of an eighty-year-old woman, from Observationum anatomico-chirurgicarum centuria *(Amsterdam, 1691).*

As stated, my fascination with Ruysch comes from his constant oscillation between his artistic and cognitive endeavors, two aspects that were not separated for him. I wanted to know how he conceived his work, and how his aesthetic motives interconnected with his cognitive purposes. Like many, I was baffled when I first saw the famous prints of the craggy fetal skeleton tableaux. When I show them to my students during my lectures, I am often asked if this man was sane. (Yes, he was!) Moreover, in contemporary sources, you only read words of praise for the way he presented his collection. The only negative comments are those of fellow anatomists who envied him.

The illustrations of the tableaux always stir a lot of wonder and reaction, but art and knowledge interconnected in more subtle ways. A good example of this is a group of stones he kept on the shelf of the eleventh cupboard. He elaborately describes how he acquired these objects. In May 1681, he was approached by an eighty-year-old woman who had been suffering for about twenty years from a protruded womb and severe pains during urinating. After examining her, Ruysch suspected that her bladder was filled with stones; during an operation, performed by two other men, the surgeons removed forty-two stones. She lived free from pain all her remaining days. Ruysch describes his wonderment at the form and structure of the stones, how they all seemed to have the same shape, even though different in size, and how smooth the surface was "as if they were polished." He considered the objects so special that he had them depicted as a tasteful seventeenth-century still life (see image above). The stones are elegantly placed on a table top or the cap of a cupboard, arranged according to their size.

Image right: Collection of stones cut from human bodies. Detail of image on page 115.

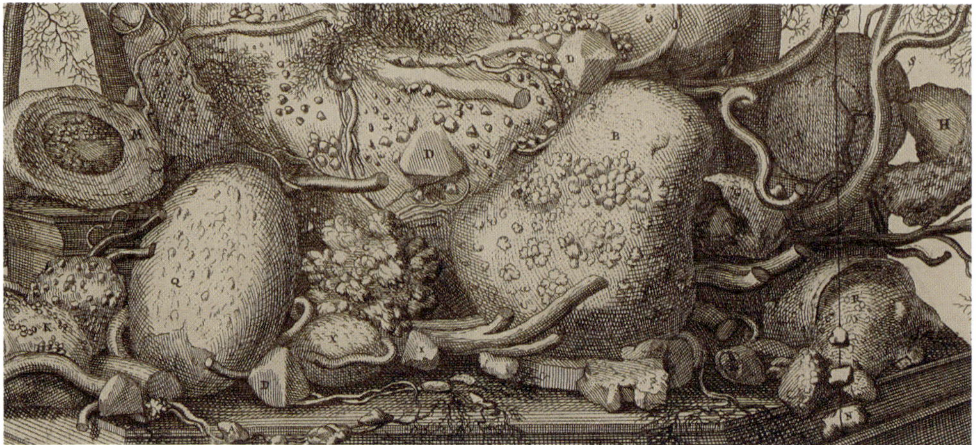

The use of sharp contrast and heavily cast shadows gives a haptic quality to these diverse, angular forms. To enhance the feeling of wonderment, the artist placed a drape at the right side, as if the stones had just been revealed. When we count the stones carefully, we notice that only forty-one are depicted, not forty-two. This is because one of them had acquired a special place in Ruysch's collection. In the third cupboard was a little box with a golden ring which was holding "instead of a diamond" one of the octagonal stones "so well polished that it was wonderful to see." Did Ruysch or one of his family members ever wear this ring? Ten years later, in 1701, the stones reappear in one of the rocky landscape tableaux he described and had depicted in the catalogue of his first cabinet, *Thesaurus anatomicus primus*. In the illustration, three of the stones are indicated with the letter D (see image above).

Many stones in the print are indicated with letters, which gave Ruysch the opportunity to provide their medical specifics and stories of origin. In this case he recounts the story of the eighty-year-old woman again. The remarkable stones serve as a good example of how objects continually changed contexts within the collection, and how Ruysch was constantly rearranging his objects to present a pleasurable, as well as an instructive, sight.

The way I depicted the collectables in this early stage of my research sometimes brings a smile to my face. For instance, the drawing marked "kastje D" (little cupboard D; see detail on facing page, top) contained an interesting object which Ruysch described as a jerkin made of human skin from a hanged thief. In Ruysch's time, convicted criminals were often used for anatomical dissections and preparations. This "garment" of human skin showed the traces of brand marks that the thief apparently bore on his skin. Back then, in 1996, I drew it quite anachronistically as a modern T-shirt, but more likely we must perceive it as a sleeveless tunic.

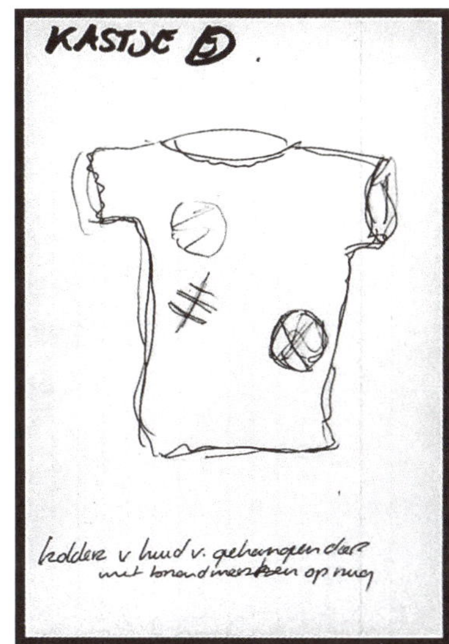

Ruysch described the same object in his catalogue of 1704 fourteen years later, after the reorganization. Apparently, by then he thought this curious object less fit for public display, because it was placed among other objects in a closed drawer below one of the larger cupboards. It was no longer visible at first sight. This transition corresponds with a general development in the history of science, in which an interest in the deviating, irregular, and spectacular made way more and more for the general, the common, and the regular in nature.

Sometimes I was able to correct my personal artistic endeavors. "Kastje G" (little cupboard G; see detail below, left) contained a "monstrous" skeleton of an unlucky deformed child that was found in October 1668 in the IJ, an inlet that formed the harbor of Amsterdam. Apparently, the poor creature was thrown in the water, probably already dead, because of its misfortune. His head was severely deformed, his torso was too long, his ribs too short, his right foot had six toes, his left foot nine, his right hand seven fingers, and his left hand six. Back then I drew it as an evil-looking fetus with a huge

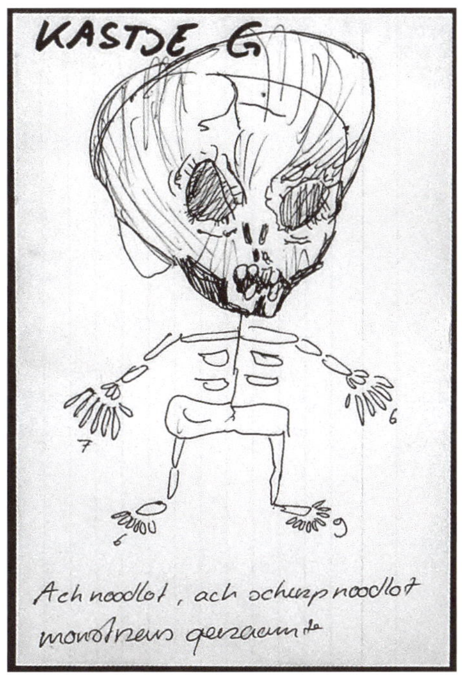

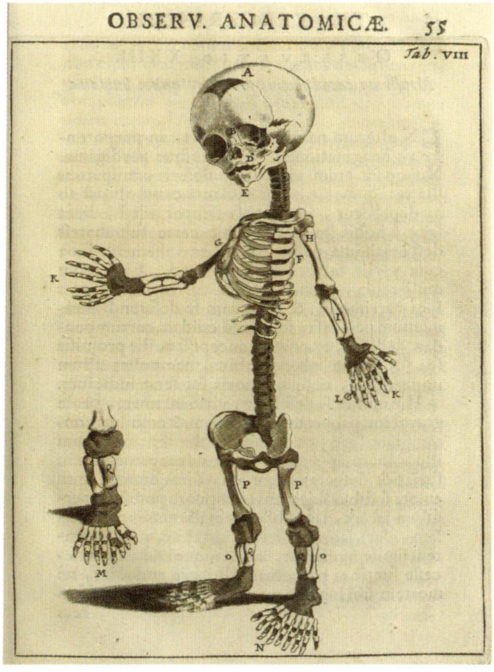

Images above and left: Details of author's sketch; image right: skeleton of a deformed baby prepared by Ruysch, from Thedor Kerckring's Specilegium anatomicum *(Amsterdam, 1670).*

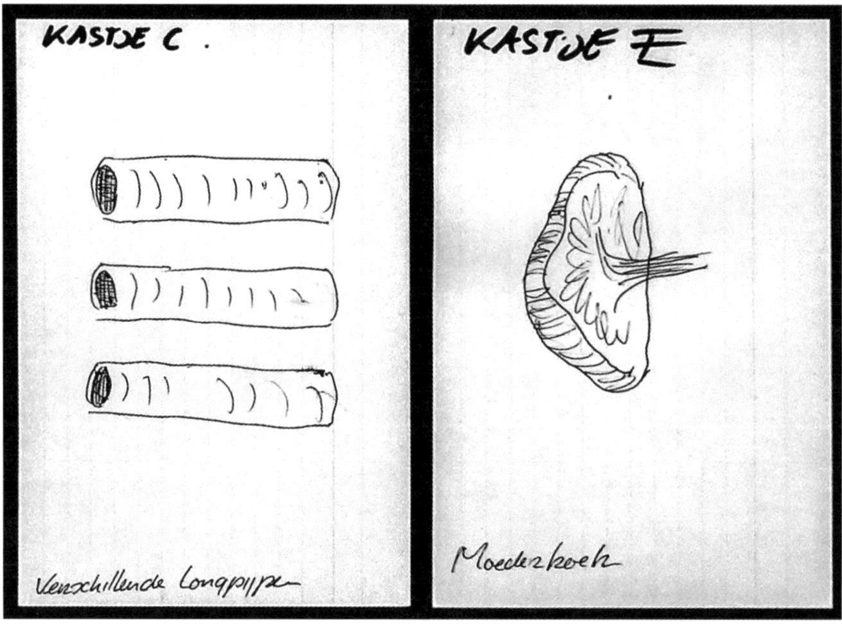

Images above and on facing page: Details of author's sketch

head. Later I found an image of this object as depicted by one of Ruysch's anatomist colleagues, Theodor Kerckring (1638–1693) (see image on previous page, bottom right). Kerckring tells how, on behalf of the city's magistrates, the dead body was brought to Ruysch, who examined it and prepared it for conservation.

In this early phase Ruysch displayed this deformed object in one of the separate little cupboards. In the later catalogues from after 1700, it is not mentioned, and he explicitly states that he was hesitant to show "monsters" like these. Instead of showing the deviant, Ruysch was now convinced that displaying the complexities and intricacy of the human body in its normal state would bestow enough wonder and amazement on the beholder. Examples of this can be found in the little cupboards labeled C, showing some bronchi, and E, showing a placenta. Ruysch saw the human body as a microcosm, the crown of the creation, made after God's likeness, magnificent in its construction. He worked hard to demonstrate the tiniest veins and the most delicate tissues in the human body to impress the beholder with its complexity.

Eight skeletons depicted in the lower rows of the two frames were not placed in the cupboards but formed part of a larger construction. In one of the rooms stood a sizable installation of skeletons and anatomical preparations. In the center was an embalmed fetus of seven months, crowned with a wreath of flowers and holding a bouquet (see image on facing page, bottom). Ruysch explains how the bouquet in its hand referred to the transitoriness of life, while the flowers in the wreath, on the contrary, referred to eternity, because the petals were perpetual: they would never wither and would always retain their

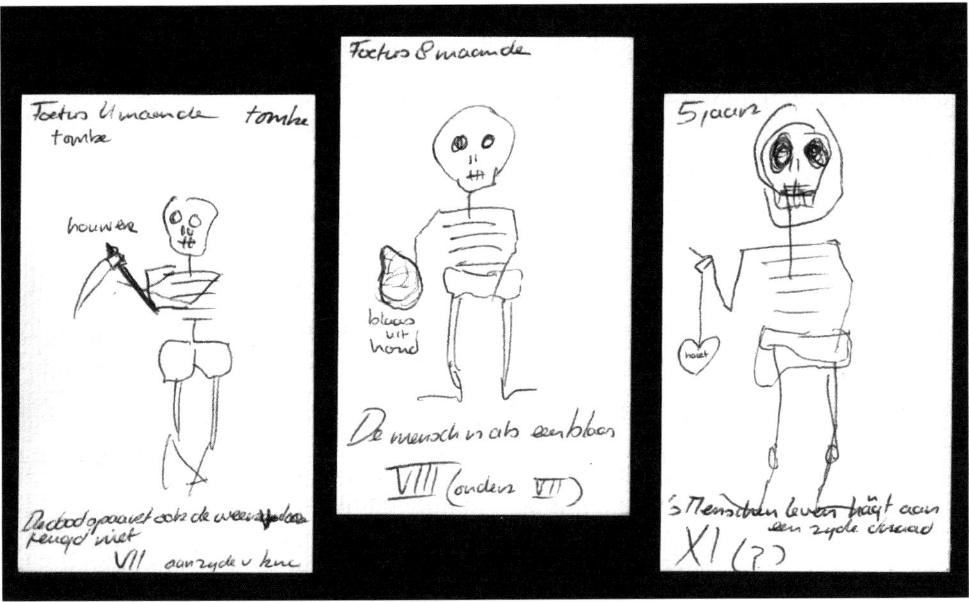

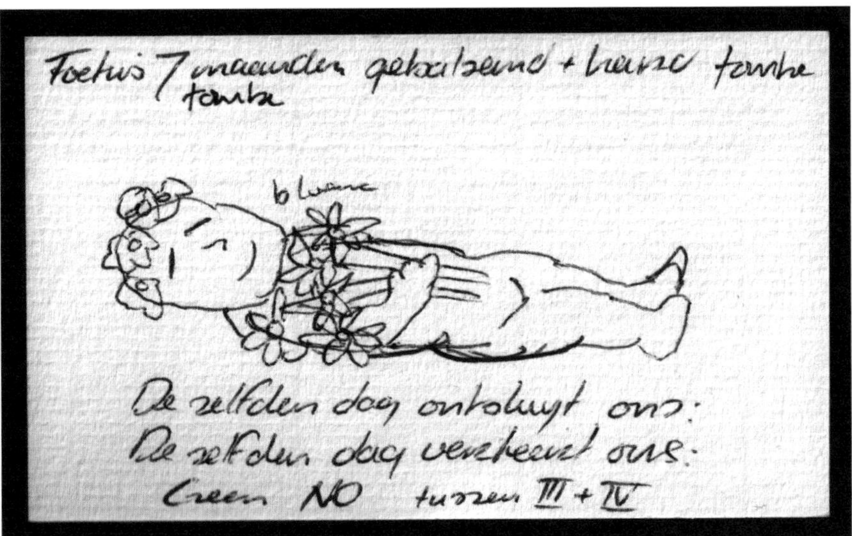

color. The embalmed fetus was surrounded by several skeletons holding symbols that reminded the beholder of the vanity of life (first frame on page 25, lower row, second and third items, and second frame, lower row, second item).

One fetus of four months was holding a scythe "threating to strike." Another fetus of eight months was holding an inflated dog's bladder expressing the famous motto "Homo bulla" (Man is a bubble). A skeleton of a newborn child was blowing a trumpet, referring to the end of days. This tomb must have been quite large, as its base was formed by a boneyard crafted from thirteen human

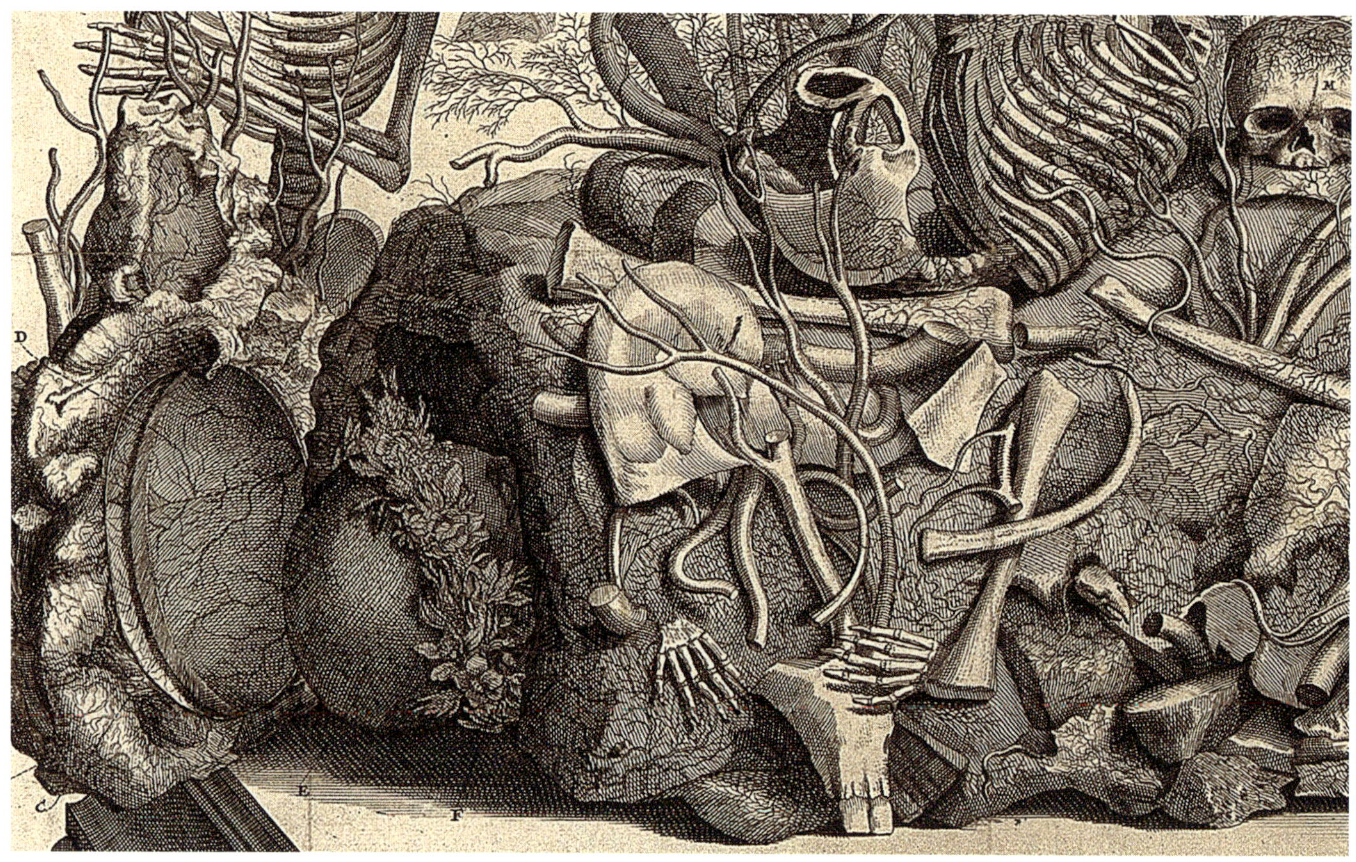

Image above: Head of an embalmed fetus with floral wreath, from Thesaurus anatomicus octavus, *1709. Detail of image on page 196.*

skulls along with many other bones. This installation was also inhabited by larger skeletons, such as that of a five-year-old child holding a heart on a thread, one of an elderly man holding his hand on his heart, and one of a child of three holding a parrot.

Through my reconstruction and my awareness of the different phases of Ruysch's collection, I could make another conclusion. As I mentioned previously, the small craggy mountain landscapes inhabited by fetal skeletons were not yet present in his early period. It is my belief that Ruysch created these artistic inventions around the turn of the eighteenth century, for a very practical reason. When he transformed his collection from dry objects to mainly wet preparations, he was confronted with a surplus of dried embalmed material. As storage space was always a problem, he found a new purpose for his older dry preparations in these artistic creations. Many of the preparations mentioned in the catalogue of 1691 reappear, incorporated in the tableaux described in the catalogues published in the first decade of the eighteenth century.

We already saw how the stones cut from an eighty-year-old woman were incorporated in one of the landscape tableaux. In a similar way, the little embalmed child decorated with flowers reappears in the catalogue of 1709 in which Ruysch described the contents of the eighth cupboard, the *Thesaurus anatomicus octavus*. Now it formed the centerpiece of another, much smaller tableau: a tomb with two crying fetuses, which Ruysch had depicted in his catalogue (see image on facing page). He explains that he constructed this tomb "artfully" from white bones, little stones, and membranes with filled veins, and describes the fetus of about six months as hard and mummified, more than twenty years old, with a crown of flowers and a bouquet in the left hand. Strangely enough the bouquet would not have been visible in the tomb. Did Ruysch take it out from this new creation when visitors came to see his collection? Again we see how he was constantly rearranging and reprocessing the objects in his collection.

The visual reconstruction helped me to better understand the development of Ruysch's collection. I decided not to visualize the second phase, from ca. 1701 to 1717, because it was not described by Ruysch at a certain point in time but, rather, over a period of seventeen years. In this new phase, Ruysch published one catalogue each year in which he described the content of one cupboard at a time, his famous *Thesauri*. Also, the number of objects was simply too vast to visually recreate. Instead, I decided to make a database of all the objects described by Ruysch in his new *Thesauri*. Ultimately, this database contained 1,125 entries and enabled me to make different kinds of calculations. I later handed the database over to the Russian-Dutch research group, who used it in their first attempts to match Ruysch's catalogue descriptions with the preparations still preserved in St. Petersburg. Recently, this group found an embalmed fetus with a wreath of flowers, but no bouquet, in St. Petersburg (see image on page 98, top). This is probably the embalmed fetus (see image on page 196) that formed the centerpiece of the installations. It is nice to know that a large part of Ruysch's legacy is still intact and attracts so much attention. How it was originally displayed by the "author" of these objects will remain a question. In 1710, he said that his collection consisted of seventeen cabinets with anatomical preparations; in addition to that, there were sixteen cabinets with animal preparations, two cabinets with fossils and shells, 1,020 boxes with insects and small animals, and 34 albums with dried plants.

When I pass the house of Frederik Ruysch at Bloemgracht 15, I often wonder what it must have looked like inside. A reliable visual reconstruction is not possible, but I can't help wondering and imagining.

The Allegorical Compositions of Frederik Ruysch

WILLEM J. MULDER

FOR TWENTY-FIVE YEARS, I WAS CURATOR AT THE ANATOMY MUSEum of the Leiden medical faculty. There I was trained by Mrs. A. M. Luyendijk-Elshout, professor in medical history, who sparked my interest in that topic. In 1991, I was appointed curator of medical collections at Utrecht's University Museum. In that period I became more involved in the European Association of Museums of the History of the Medical Sciences. Here I got to know Anna Radziun, curator of the anatomy collections at St. Petersburg's Kunstkamera. She asked me if I might be able to assist in the restoration of some specimens made by Frederik Ruysch that were part of her collection, as she knew that I had worked on some of his specimens in our own collection at the Leiden Anatomical Museum.

For this restoration project, which stretched over 10 years, we agreed that it was important to preserve the integrity of the objects by modifying the original specimens as little as possible. We refreshed the alcohol only when necessary, and kept the dissections in the original jars. Once completed, we sealed them with a special kit that created a sort of hermetically sealed chamber.

Just after the start of the renovation project, I discovered *Alle de ontleedgenees- en heelkundige werken...*, a posthumous edition of Ruysch's collected works. Intrigued with what I saw, after my retirement I transcribed the entire book, a major work that took me over a year. In this way, I gathered some expertise in the works and life of this famous anatomist. I also discovered that there were several specimens in St. Petersburg that are neatly described in the book. Then I was part of a group of experts (the so called "Ruysch Group") that produced the volume *Geloof alleen je eigen ogen* (Believe your own eyes only) describing Ruysch's productions, published in 2017.

The most intriguing aspect of Ruysch is, no doubt, his three-dimensional allegorical compositions, still lives crafted from human, animal, and botanical materials. At one time, they were on display in his home in the Amsterdam Bloemgracht, where he kept all his anatomy specimens. Ruysch was proud to give guided tours for distinguished foreign visitors, colleagues, and students. Other interested people were also welcome, but had to pay an admission fee. The tableaux are believed no longer to exist, but they have a certain immortality in the spectacular engravings of Cornelius Huijberts (1669–c. 1712), a skilled and distinguished engraver who created a great many works for Ruysch.

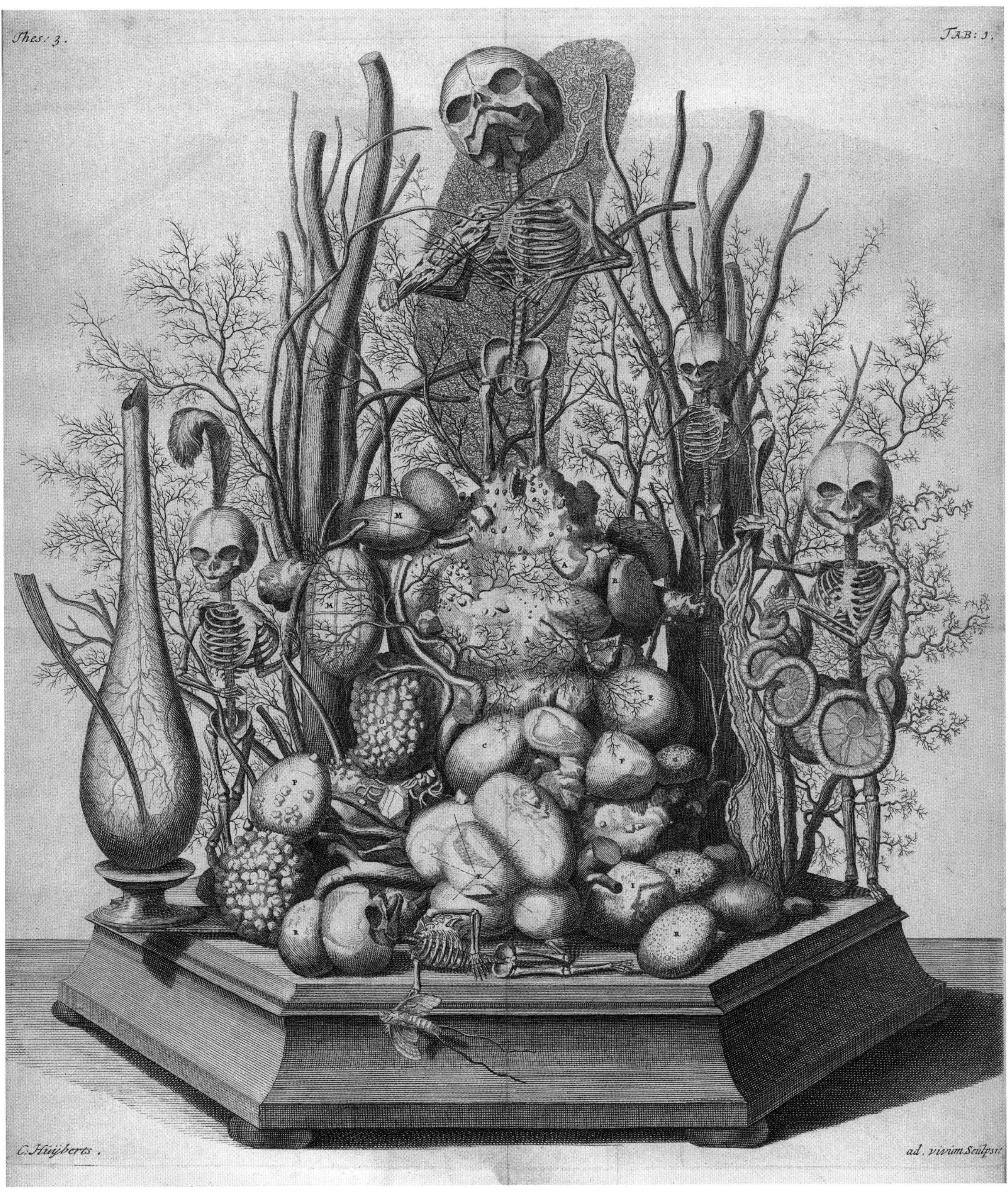

Thes: 3.
TAB: 1.
C. Huijberts.
ad. vivum Sculpsit

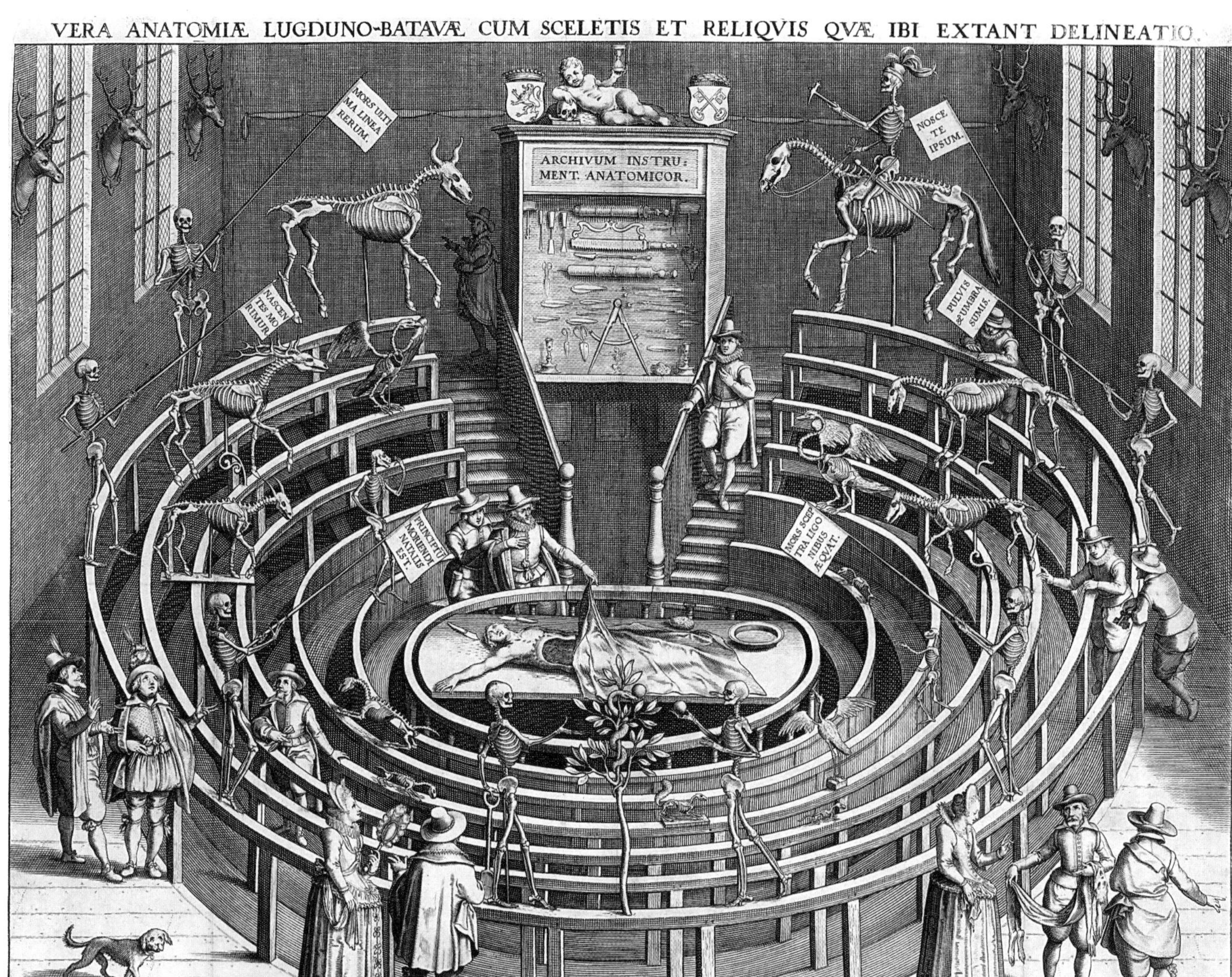

Image on previous page: One of Frederik Ruysch's famed allegorical compositions; for another, *see page 115.*

Ruysch described his compositions in his *Thesauri*, the catalogues in which he listed all his specimens and collections. He was in a position to collect many of these specimens due to his professional roles as an anatomy professor in the Athenaeum Illustre in the city of Amsterdam, where he was also in charge of the training and examination of midwives as well as an official of the medical police and coroner, engaged by the city burgomaster.

The tableaux are more or less surrealistic living pictures that attract the viewer with astonishing surprise, as well as eliciting tenderness and even some alarm. One has to keep in mind that Ruysch approached his profession with a respectful mind to God and Nature, but at the same time he demonstrated and expressed his own artistic view.

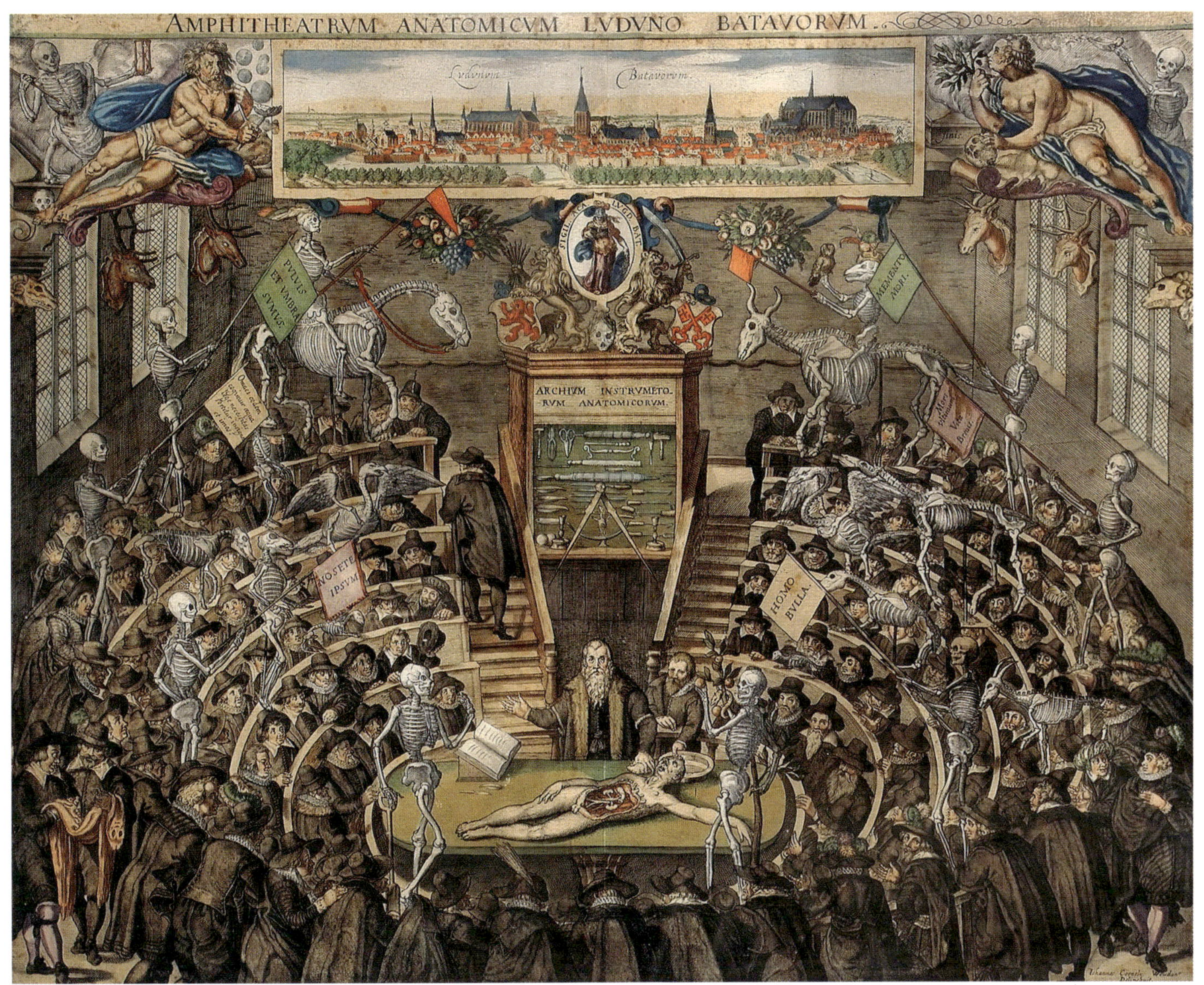

THE COMPOSITIONS AS SUCH

In Roman Catholic countries, churches were routinely decorated with *danse macabre* frescoes, in which animated skeletons and corpses led people of all social classes on a merry dance to the grave. After the iconoclasm of the Reformation reached Holland (1566), Protestant churches became more sober, but Ruysch, being a Reformed Christian, could utilize the *danse macabre* themes in his anatomy cabinets, devoid of their original Catholic connotations. His original compositions were assembled with more or less artificial rock formations prominently decorated with tiny skeletons of babies and fetuses surrounded by an arrangement of human kidney-, bladder-, and gallstones, corals, insects, plants, and wax-filled human or animal blood vessels.

Two views of Leiden's anatomical theater. On facing page: an engraving dated 1610 by Willem Swanenburgh. Above: a color depiction in the collection of the Museum of Normal Anatomy of the Military Medical Academy in St. Petersburg.

With his compositions, Ruysch aimed to express the vulnerability of life. He was specifically inspired by the anatomical theater of Leiden, the town where he studied medicine. It was once installed in the Faliede Bagijnkerk, a former church. Here, a number of human and animal skeletons were mounted on top of the circular galleries. Some of these skeletons held flags bearing mottos such as "*Vita quid est? Fumus fugiens et bulla caduca*" (What is life? A transient smoke and fragile bubble) and "*Memento mori*" (Remember you will die).

In the late twentieth century, a modern recreation of this Anatomy Theater was built at the Museum Boerhaave, the National State Museum of Science and Medicine, also in Leiden.

Ruysch, who originally was educated as a pharmacist and botanist, assembled, dissected, and collected numerous human anatomical specimens, after his medical training and for the rest of his long life. He was especially interested in the blood vessels but also in teratology, the study of inherited or other malformations. During the many years when he was instructor in anatomy and obstetrics, he enlarged his collection and gladly welcomed visitors.

One of the most illustrious of these visitors was tsar Peter the Great, who visited while in Amsterdam and Zaandam to study shipbuilding. Peter was most impressed by the collection of Ruysch; so interested that he purchased the entire anatomical collection, which at that point contained over 2,000 objects, probably including the allegorical compositions. The transport of the collection to Russia by ship must have been no easy task. However, there is some evidence that all specimens but one reached St. Petersburg safely. After this, Ruysch, with unremitting energy, began to build a new collection. It was sold at auction after his death in 1733.

SHIPPED OR NOT?

What happened to Ruysch's great allegorical compositions is not exactly clear. Early visitors to the Kunstkamera, the museum specially erected by Peter the Great for the education of the common Russian people, never mentioned their existence. In 1747, a great fire destroyed a part of the Kunstkamera, though luckily the anatomical collection escaped destruction. It is also possible that Ruysch's compositions were destroyed because they were thought to be profane. Karl Ernst von Baer (1792–1876), curator of the collection in the early nineteenth century, never registered the compositions. Also, V. V. Ginzburg, who produced a new catalogue of the collection in 1947, was not familiar with these impressive objects, and concluded that they must never have made it to the Kunstkamera.

The Kunstkamera, now known as the Peter the Great Museum of Anthropology and Ethnography, still houses some 900 pieces of Ruyschiana. We do not know for sure whether von Baer and Ginzburg were correct in their belief

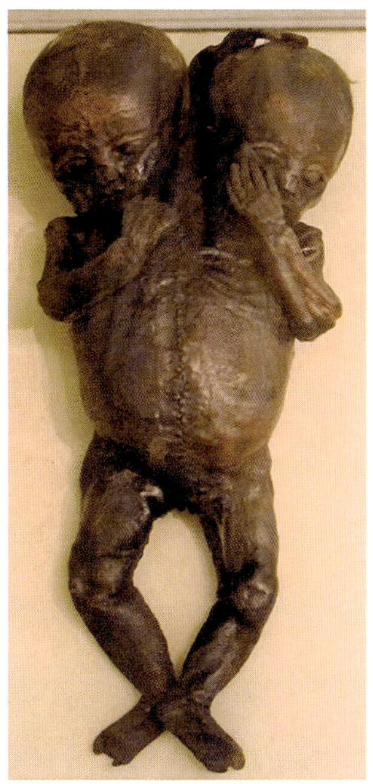

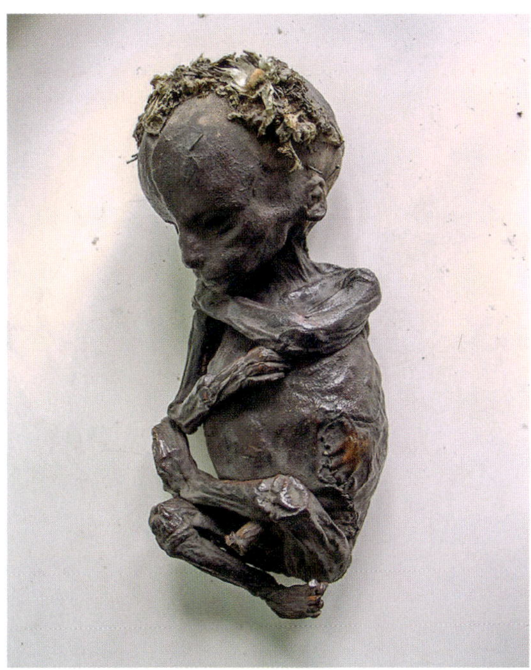

that the allegorical compositions never reached St. Petersburg. It would not be surprising if they had been destroyed, for we know they were very fragile. Ruysch was proud of the fact that he erected these skeletons on their own bones, using only the dried bones, natural ligaments, and membranes to mount the several parts. He did not use any iron or wooden supports. The result must have been very beautiful, but at the same time extremely delicate.

AN INTERESTING DISCOVERY

During the restoration project on the collection in the Kunstkamera we discovered, in an old shoe box in the museum storage area, a small mummified female fetus of about six months. A closer look made us believe it might be a fetus Ruysch described in the tableau in his eighth cabinet. We compared Ruysch's description in *Alle de werken* to the piece, and were quickly convinced that it must be the same. This provides at least some evidence that the allegorical compositions did actually make their way to St. Petersburg.

Image left: The conjoined baby twins described in Alle de werken.

Image right: The mummified fetus decorated with flowers on her head.

ANOTHER COMPOSITION

Ruysch described his fourth composition in *Alle de werken*, unfortunately without an illustration. We find the description in his seventh cabinet: "On the ground you find a grave tomb made of several East and West Indian horns of snails, shells, and sea branches. This tomb held a two-headed fully grown baby that has three arms and two legs. I bought this years ago from others who embalmed and hardened it; it is a pity this is not of the quality that I strived for. Its color is not as good as natural." This specimen is still in the collection of the Kunstkamera: inventory number 4070-913.

CONCLUSIONS

The find of the "girl with the corolla of flowers" and the conjoined twins in the storage of the Kunstkamera in St. Petersburg makes it likely that the great allegorical tableaux were packed with the rest of the collection. What is sure is that Ruysch exhibited great virtuosity in his presentations of his findings and discoveries. His aesthetic and artistic—as well as scientific—approach compels not only respect but also admiration. And, although we can no longer see his great allegorical tableaux, many of his rightfully famous specimens are still to be seen in the St. Petersburg Kunstkamera, as well as a few in the Netherlands.

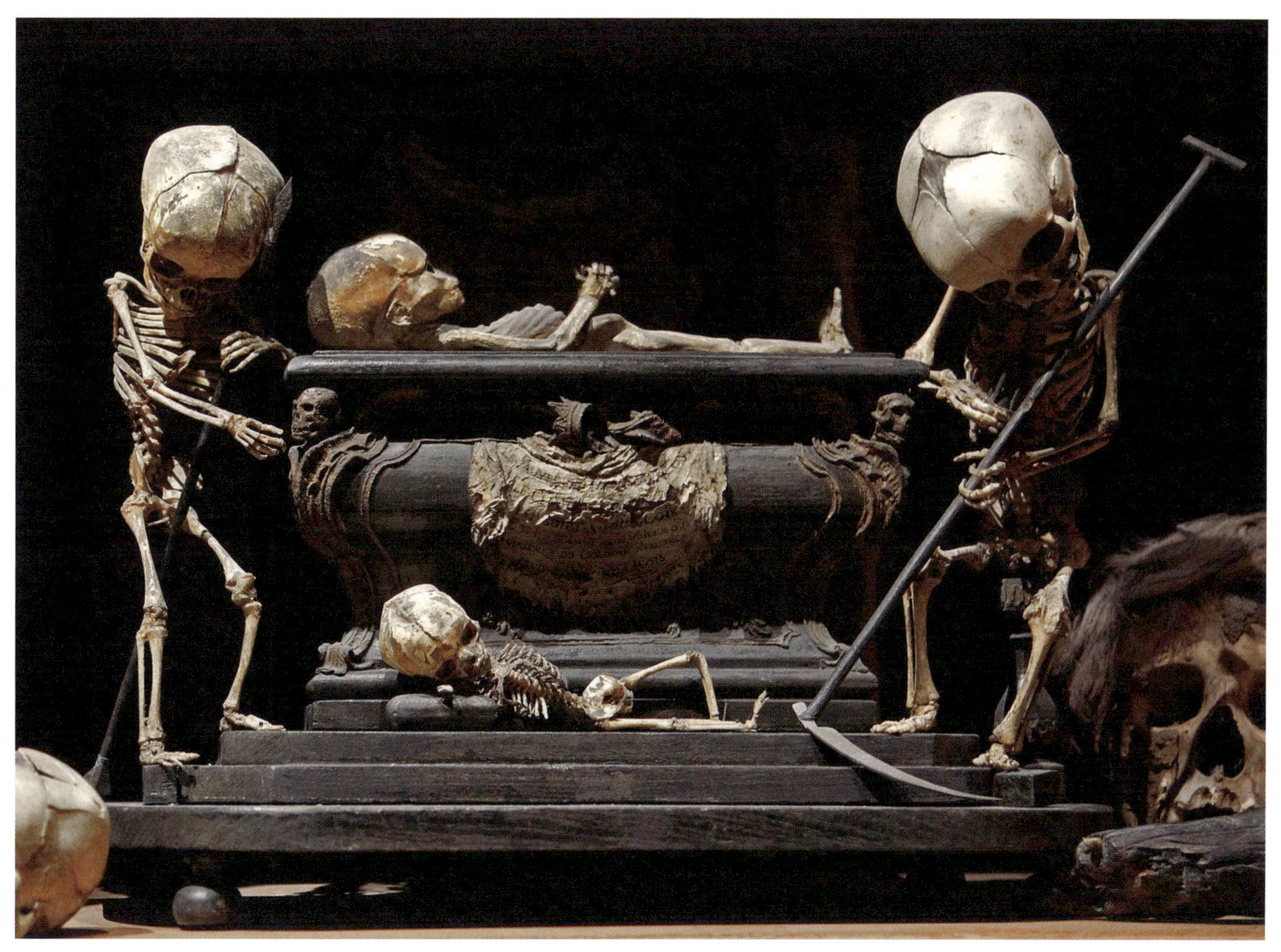

The Macabre Altar: Sole Survivor of the Sue Collection

PHILIPPE COMAR

AROUND 1745, SURGEON AND ANATOMIST JEAN-JOSEPH SUE THE ELDER (1710–1792) became assistant professor at Paris's Académie Royale de Peinture et de Sculpture. Around this same time, he began collecting the items that, by the French Revolution, would become one of Europe's greatest anatomical collections. A master in the art of dissection, he invented new procedures for preparing and preserving anatomical specimens. In 1748 he published his *Abrégé de l'anatomie du corps de l'homme, avec une méthode courte & exacte sur la manière d'injecter & de préparer les parties fraîches ou sèches*, followed by a work titled *De la manière d'ouvrir les cadavres humains, et*

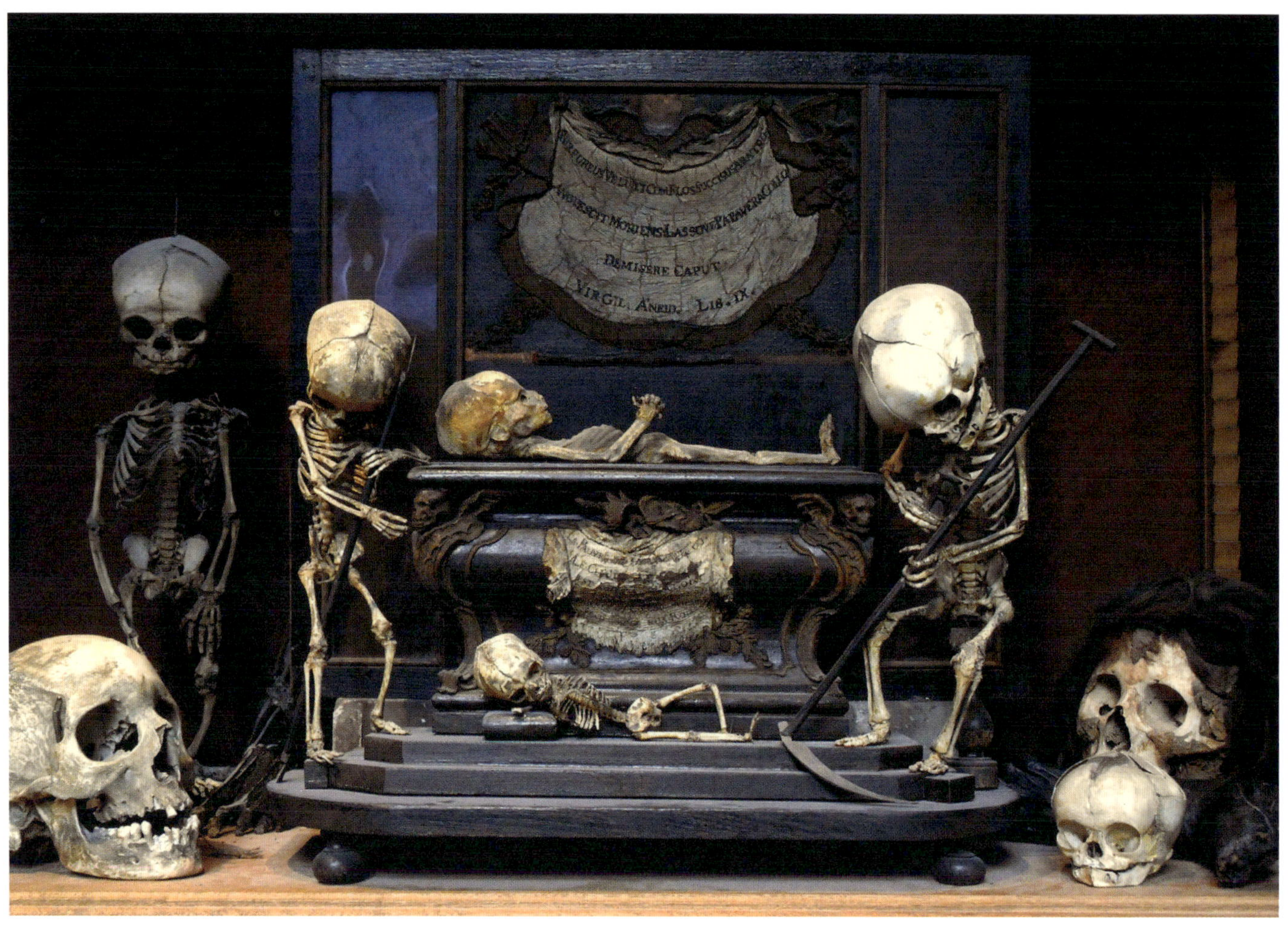

Above and previous page: Macabre altar with a mummy and three human fetus skeletons. Anonymous, late seventeenth century; skeletons, mummy, wood, metal, glass; 49 x 56 x 24 cm. École des Beaux-Arts, Paris, MU 12243. Above: photo by Philippe Comar; previous page: photo by Joanna Ebenstein.

de les embaumer, a work he republished a few years later under the generic title of *Anthropotomie*, in other words the art of "cutting people up."

His nephew, Pierre Sue, who was also a surgeon, wrote that "for M. Sue, the amphitheater in the Collège de Chirurgie had become a kind of palace where he displayed all the treasures of his anatomical works."[1] His anatomical preparations had been famous since the mid-eighteenth century, not just because they demonstrated a triumph over the corruption of the flesh, but also because of the power they had over the imagination. The article titled "Tanned Human Skin" in Diderot and d'Alembert's *Encyclopédie* describes one such preparation: "M. Sue, a Paris surgeon, presented to the Cabinet du Roi [the royal natural history collection] a pair of slippers made of *human skin*, preserved in such a fashion that the hairs on this *skin* were not damaged in any way." This pair of slippers would earn Sue the reputation of being "the man who walked in others' footsteps."[2]

In 1785, at the age of 75, Jean-Joseph Sue the Elder passed on his position of surgeon at the Hôpital de la Charité to his son, also named Jean-Joseph (1760–1830). Most importantly, however, he entrusted him with the care of his collection, which at that time comprised 120 anatomical preparations and 195 plates, some of which were in color. After just a few years of his son's management, the collection swelled, growing to 1,300 items and 364 plates.[3] Inspired by this expansion, Jean-Joseph Sue the Younger created in his father's home, where he lived, a cabinet that was open to the public and sufficiently well-respected that in 1787 the *Guide des amateurs et des étrangers voyageurs à Paris* considered it worth a visit: "M. Sue the Younger ... , living in Rue des Fossés-Saint-Germain-l'Auxerrois on the corner of Rue de l'Arbre-sec, has assembled a valuable collection of all that might pique the curiosity of a connoisseur in this field."[4]

In 1792, after his father's death, Jean-Joseph Sue the Younger succeeded him as Chair of Anatomy at the Académie Royale de Peinture et de Sculpture. Around 1805, he moved to the district near Boulevard de la Madeleine, where he set up his cabinet of anatomy again. It was here, under the Restoration, that he gave lectures to a select group of people: "On Thursday April 29, M. Sue will launch a series of lectures in physics—that is, the natural sciences—specially tailored for the general public. Please apply to M. Sue, 3 Rue du Chemin-du-Rempart, every day between midday and two o'clock."[5]

In February 1824, the anatomy lectures of the École des Beaux-Arts (which had replaced the Académie Royale de Peinture et de Sculpture) acquired a new lecture room in the former Musée des Monuments Français, where the École had taken up residence. Sue the Younger, whose house was threatened with demolition because of development works in the Madeleine district, offered to transfer his cabinet of anatomy to the École for a period of fifteen years.[6] On November 13, 1824, the Interior Minister welcomed this loan: "I applaud M. Jean-Joseph Sue's generous offer and I am sure that great educational benefits will result from the presence in the anatomy lecture hall itself of a collection that will give the teacher the means and many opportunities of expanding more fully on the themes of his lectures."[7]

In 1825, in *Extrait du discours prononcé par M. le professeur Sue ...* , an editorial note stated that the Sue collection had long enjoyed a reputation throughout Europe and that foreign countries had even offered to buy it, but that Sue the Younger had "constantly refused these wonderful offers, preferring to keep it in his own country." Professor Sue told his audience:

> *In this museum we see a collection of ruins that were once alive, authentic medals of the natural history of man; relics snatched from the second death that awaited them in the tomb; precious remains which, like the imposing columns of an ancient palace destroyed by the hand of time, awaken great ideas and command a kind of veneration. ... Who can count the wonders that arouse and vie for the viewer's attention? The pathways of blood high-*

> *lighted by a colored fluid, lymph replaced by mercury, branching structures that one might be tempted to think were the most unbridled creations of the plant kingdom, the nervous system, an astonishing kind of tree whose roots occupy the upper parts and branches the lower part; human brains laying bare the workshop once home to thoughts and forming the first element in a progressive series comprising the brains of all the animals according to their level of intelligence; … the flexibility of viscera preserved by the fluid in which they are immersed; the shades of color and freshness restored, in a way, to the most delicate organs; nature celebrated in all shapes and sizes; destruction averted as much as it can be; death finally rejoicing in a kind of life, or at least in an outward display that better suits our delicate sensibilities; everything in this philosophical museum, in this library of relics and bones, stimulates the soul, interests it, stirs it, and shakes it up.*[8]

Image on facing page: Illustration from Andreas Vesalius's De humani corporis fabrica, *1543. Wellcome Collection.*

The writer Alexandre Dumas (1802–1870) noted that the collection also included "Mirabeau's brain."[9] In March 1829, at the age of 69, Jean-Joseph Sue announced his intention to make a "complete and definitive gift of the rich collection of anatomical models which he had temporarily donated, and which has been stored since that time in the large hall assigned to the anatomy lectures."[10] The permanent donation stated that a "catalogue" was included with the deed of gift, but of this only the first manuscript page remains, which includes just the first twelve inventory numbers among some 1,600 items.[11]

Jean-Joseph Sue the Younger died the following year. In 1833, a note from the École des Beaux-Arts drew attention to the "deterioration of the Sue anatomy collection."[12] A year later, following a visit by the government minister responsible, a report was sent to him about the cause of the damage to Sue's cabinet of anatomy:

> *Monsieur le Ministre, the problem remains of how to account for the immediate cause of the foul air that you identified in the cabinet of anatomy. This valuable collection was created more than 50 years ago with great care by Monsieur Sue, a member of the former Académie de Peinture and teacher at the École. It was passed on to his son who … decided to donate his collection to the École, in the hope that, once accommodated in a favorably appointed room, it might be useful to the students of the École. This has not happened. … In the nine years since the cabinet was set up here, decomposition of all the items it contains has set in with deplorable speed. The amount of turpentine and spirit-based varnish used to protect against mildew has accelerated the decomposition of the injections, and the resulting deliquescence has created the foul odor emitted by the cases. Having no means of remedying the total decomposition of the most delicate parts of this cabinet of osteology,*

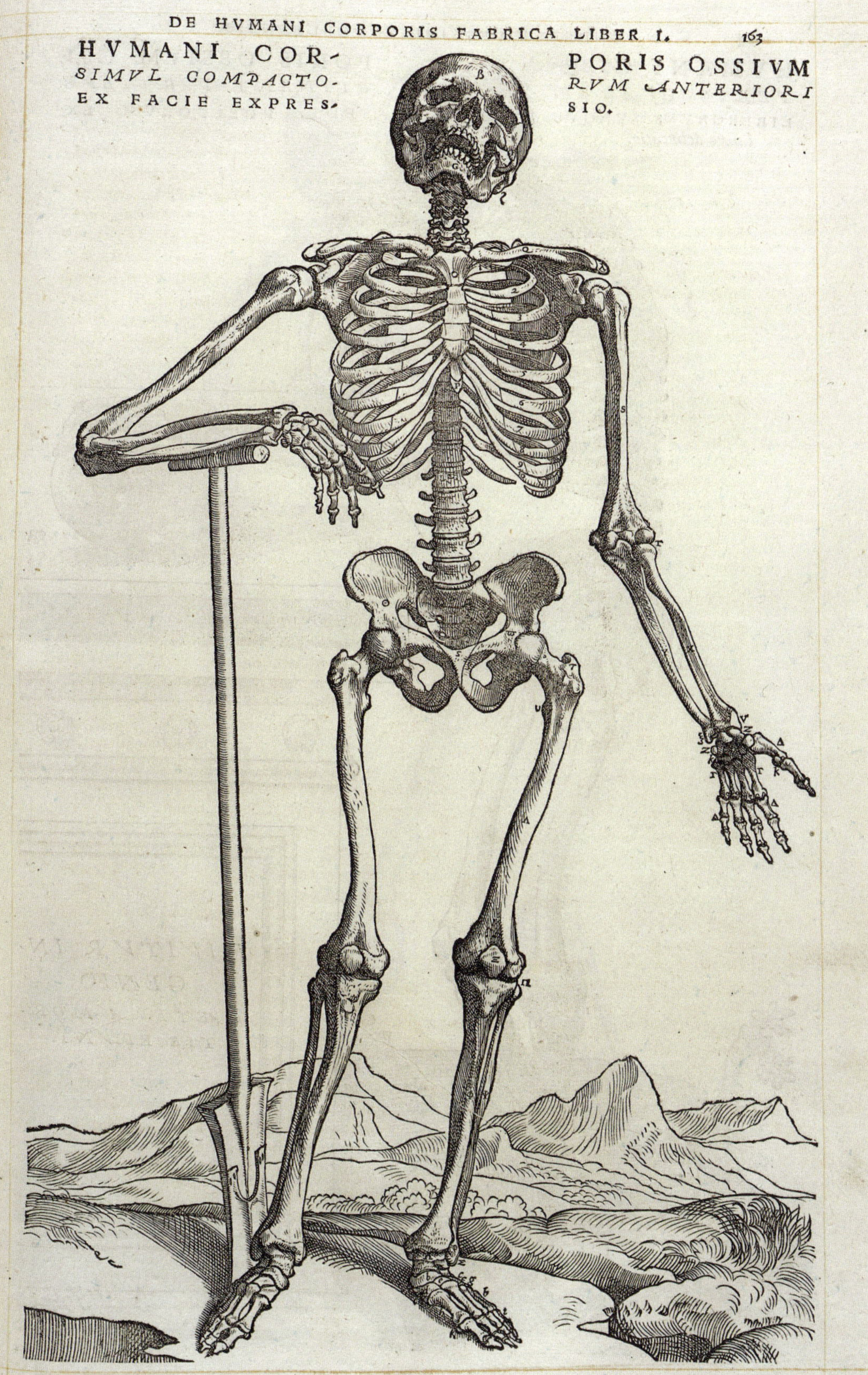
DE HVMANI CORPORIS FABRICA LIBER I. 163
HVMANI COR-
SIMVL COMPACTO-
EX FACIE EXPRES-
PORIS OSSIVM
RVM ANTERIORI
SIO.

> *as well as the ruin of the apparently most robust parts which are attacked by the destructive action of saltpeter, the administration resigned itself with the deepest regret to the loss of a collection which it had received in an already substantially damaged condition. … Such is the background to the destruction of a very rare and very valuable collection.*[13]

All seemed definitively lost. Yet twenty-four years later, one item from Sue's cabinet resurfaced. In December 1858, the director of the École asked a man named Lintilhac to return the "skeleton of the Flute Player in Sue's cabinet,"[14] which had been entrusted to him by M. Robert, the professor of anatomy. So it seems Sue's cabinet included a flute-playing automaton, or at least what might remain of it. When did this android musician enter the École's collections? Probably in the second half of the eighteenth century, when such automata were very fashionable. We are surely reminded of the famous *Flute Player* built in 1738 by Jacques de Vaucanson (1709–1782), an engineering genius who was passionate about anatomy and an exact contemporary of Jean-Joseph Sue the Elder. It is not known if the object was indeed returned. In any case, it is no longer in the École's collections.

Have other items been recovered? Apparently, all those that were likely to decay were destroyed around 1835. It is possible that some dry preparations that were less susceptible to putrefaction, such as fragments of mummies or skeletons, escaped destruction. But these are just hypotheses. What is known is that one of the oldest items preserved in the Galerie Anatomique of the École des Beaux-Arts—and also one of the most extraordinary—seems to have come from the Sue cabinet. This is a late seventeenth-century macabre altar, which comprises a mummy and three skeletons of human fetuses. In 1795, an inventory of natural and artificial anatomical items in the Cabinet d'Histoire Naturelle of the Hospice de l'Unité (formerly Hôpital de la Charité), where Jean-Joseph Sue the Younger worked as a surgeon, mentions a "kind of allegorical tomb under glass, incorporating four skeletons of fetuses from different periods."[15]

This memento mori is arguably the only item that has survived this collection's destruction—an ironic fate for the fabulous Sue collection. This secular reliquary comprises a baroque cenotaph decorated with skulls bearing bats' wings. In Homer's *Odyssey*, bats represent the souls of the dead (Book XXIV). On the altar there lies the mummy of a human fetus of about four months. The drapery above transcribes Virgil's lines on the death of Euryalus: "Purpureus veluti cum flos succisus aratro / languescit moriens; lassove papavera collo / demisere caput / Virgil, Aneid. Lib. IX" (Like a bright flower, scythed by the plow, bowing as it dies, or a poppy … bending its weary head). The inscription that adorns the bottom of the altar reproduces verses by a French poet, François de Malherbe (1555–1628), from his "Consolation à Monsieur du Périer sur la mort de sa fille," about a child who lived for "the space of a morning": "The poor man in his hut, beneath the thatch, is subject to [Death's] laws, and the guardsman who protects the gates of the Louvre cannot defend our Kings from it." The cenotaph is flanked by two skeletons of fetuses about six months old, standing upright. Their skulls, drooping onto their sternums, seem to illustrate the quotation from Virgil: "Lassove papavera collo demisere caput," as the head of the poppy suggests a tiny skull. The skeleton on the left holds a feathered pen to write the names of the deceased in the Book of the Dead. The one on the right carries an hourglass and a scythe, two recurrent emblems in the Christian iconography of death, and which are originally attributes of Saturn. The latter was said to control weights and measures and castrated his father with a sickle. Saturn, the equivalent of the Greek Cronus who devoured his children, was incorporated into the god of death very early on. He mutilated his father and at the same time cut off his lineage.

This memento mori inevitably recalls the preparations made by the doctor Frederik Ruysch, who was employed to teach midwives in Amsterdam. Ruysch used the corpses of stillborn babies in his dissections, and from 1679, as a forensic physician in the city, made use of the bodies of infants drowned in the harbor by their parents. Indeed, *The Anatomy Lesson of Dr. Frederik Ruysch*, painted by Jan van Neck, shows him dissecting a newborn baby. Thanks to his talents as a chemist, he perfected the processes of preservation and embalming, and created a cabinet of anatomy where he displayed his preparations. Some of them were almost like three-dimensional paintings, as illustrated in the engravings in his *Thesaurus anatomicus*, featuring baroque settings of fetal skeletons living among treelike networks of arteries, anatomical waste, gallstones, and kidneys. Animal tissues give the illusion of plant and mineral forms, designed to create a fantastical landscape where fetal skeletons, lost in the middle of a forest of injected bronchi, dry their improbable tears using handkerchiefs cut out of a piece of mesentery. These anatomical curiosities, evoking in equal measure ancient child sacrifices and a macabre version of a Charles Perrault fairy tale, turn Frederik Ruysch into a saturnine, ogre-like figure.

In the memento mori in the Galerie Anatomique of the École des Beaux-Arts, the fetal skeletons seem less like victims than agents of death. The macabre altar is a vanity, but far from reminding man that he is destined for the void, its setting reminds him that there is, beyond death, another existence that has its own rules, rites, and inscriptions, and that the skeletons are only seemingly dead as they are still endowed with the kind of life that lets them move about, cut down the living, and write their names in the Book of the Dead. These creatures exist in a world where time's arrow has disappeared: they are dead before they were born. It is a magnificent short circuit. They escape the ephemeral world of the living to live, indifferent and disdainful, in the eternal world of the dead.

In the plates of Andreas Vesalius's revolutionary anatomical treatise *De humani corporis fabrica*, published in 1543, we see a skeleton vigorously digging the ground, suggesting that there is life after death. The promised repose does not exist. In the century of Charles Darwin and materialism, by contrast, there is nothing after death. Each of us is summoned to return to the void. Death, with all its theatricality, its draperies and altars, its dancing skeletons and its bats, with its ghouls and voices from beyond the grave, is dead—and has died a natural death.

In Giacomo Leopardi's "Dialogue between Frederick Ruysch and His Mummies" of 1827, the conversation with the dead ends as follows: "Let me examine them a little. Yes, they are quite dead again. There is no fear that they will give me such another shock. I will go to bed."

Enchanted Anatomists: From Frederik Ruysch to Louis Bolk

LAURENS DE ROOY

WHEN FREDERIK RUYSCH SOLD HIS FAMOUS FIRST COLLECTION OF anatomical preparations to tsar Peter the Great in 1717, he promptly started work on a new collection. After Ruysch died in 1731, most of this second collection was sold to August the Strong, king of Poland and elector of Saxony (1670–1733). Smaller portions were sold to others and stayed in the Netherlands, but none of these remained in Amsterdam. For many this is an eye-opener: the museum I run, Amsterdam's anatomical museum, the Museum Vrolik, houses a large collection of anatomical treasures, around 10,000 preparations, skeletons, and skulls of humans and animals, but none are from Ruysch; they all date from after his death.

If not his actual collection, what, then, is the material and immaterial legacy of Ruysch in Amsterdam? Of course there are the paintings: his portraits and anatomical lessons. However, perhaps his most important material legacy is a building. For, as praelector of anatomy and surgery for Amsterdam's Surgeons' Guild, Ruysch supervised the construction of a new anatomical theater. It was erected in a newly built central tower in the Waag (weighing house) on Nieuwmarkt square and put into use in 1691. Because of its function, the tower would become known as the Snyburg ("carving tower"). In this theater and adjacent rooms, the Amsterdam anatomists would perform their dissections for almost 200 years. The surgeons' coats of arms were painted on the ceiling of the theater. In the center of these appears that of Ruysch himself. The ceiling remains untouched to this day. It is the only visible remainder of Ruysch as creator of this magnificent space.

But is there also a kind of immaterial legacy? Some sort of Amsterdam anatomical tradition, in the sense that we see elements of Ruysch's research style, preparation methods, and philosophy in his successors? For that I will take a bird's-eye view of a number of Amsterdam anatomists in the two centuries after Ruysch's death. What was their philosophy of life, and how did it influence their research methods? Every now and then I will also consider the workshops where anatomical knowledge was produced: the theater, the laboratory, the museum.

RUYSCH AND THE MECHANICAL WORLD

What were Ruysch's ambitions, and what was his approach to life? I will only provide a short recap here, for most of it is discussed in other chapters of this book.

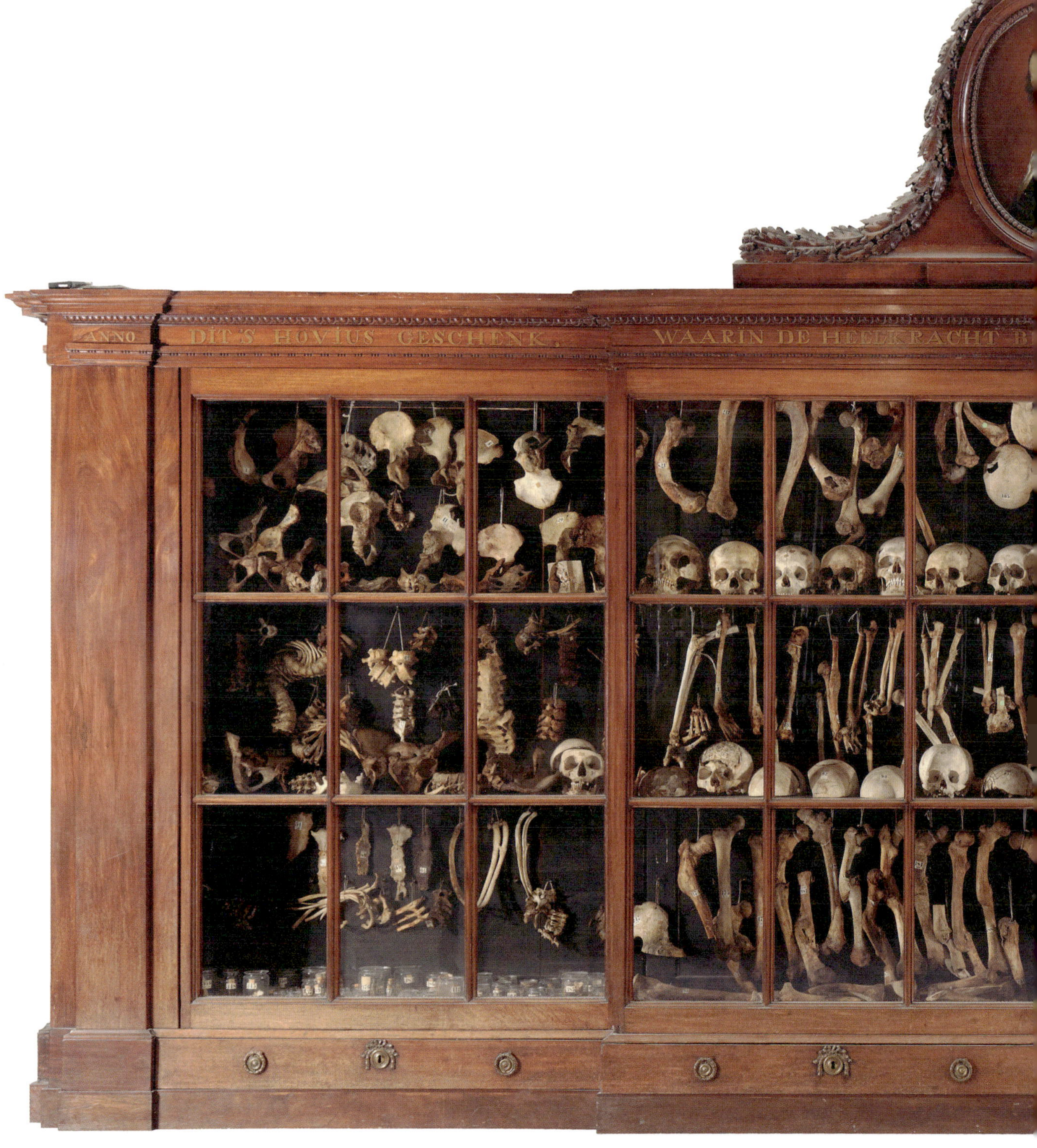
ANNO
DIT'S HOVIUS GESCHENK,
WAARIN DE HEELKRACHT

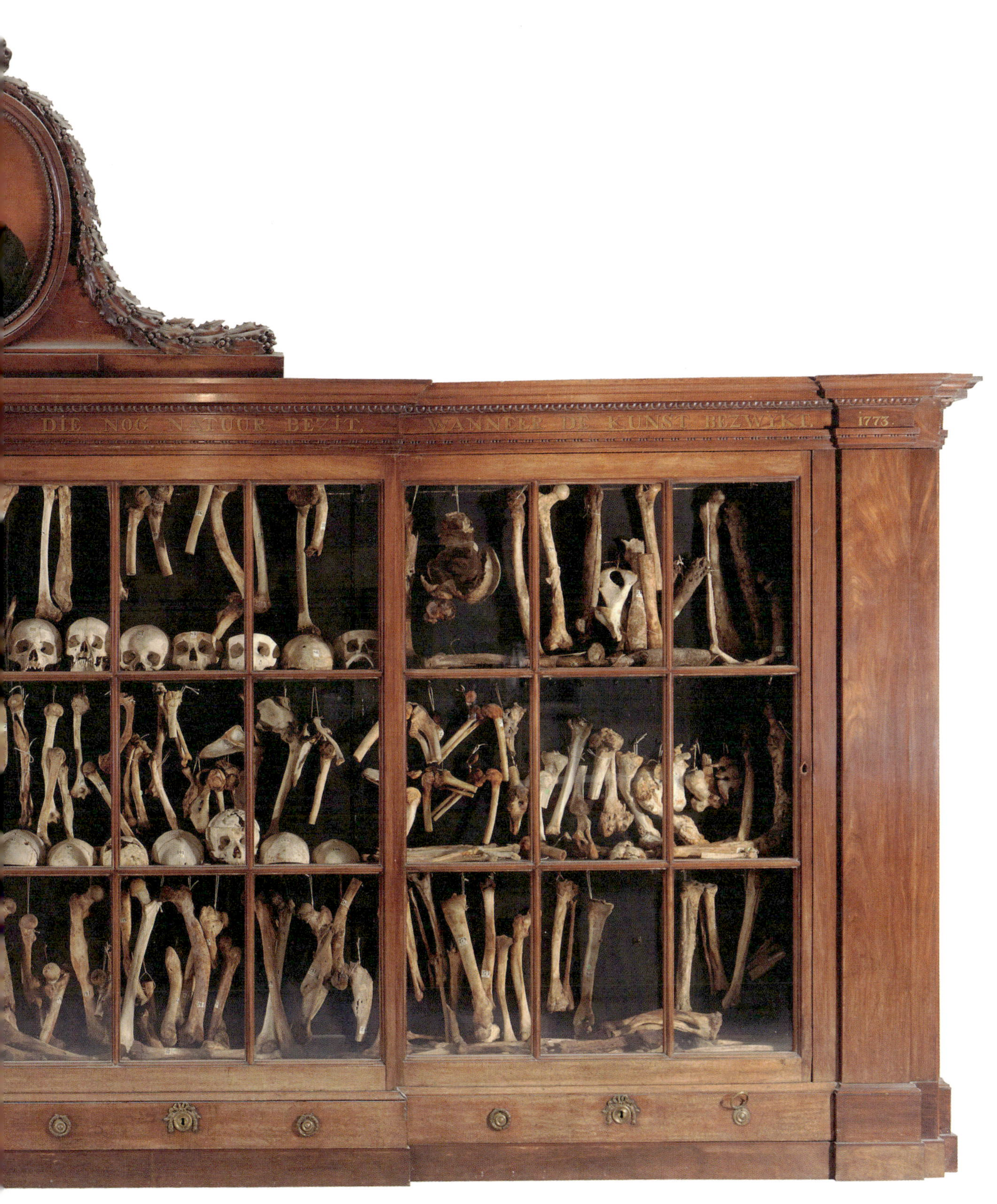
DIE NOG NATUUR BEZIT. WANNEER DE KUNST BEZWIKT. 1773.

Previous pages: Hovius's cabinet from Museum Vrolik, the Amsterdam Museum of Anatomy. The cabinet dates from 1773. Photo by Hans van den Bogaard (2009).

Both in his museum and his publications, Ruysch paired his preparations with Latin memento mori quotations by Roman writers in order to avoid Christian references and ideas surrounding mortality and death. With his preparation and injection techniques, he was able to reach the thinnest capillaries of the skin in such a way that some of his preparations seemed to defy mortality. His motivation for creating these preparations was to evoke emotion, to urge the viewer to contemplate mortality and to make these dead parts less frightening.[16]

In his anatomical descriptions, on the other hand, he did not refrain from emphasizing how the human body was created so beautifully by the Almighty, and how privileged he felt that it was he who had been allowed to discover it all. He asked poets and doctors who were his friends to write poems as introductions to his anatomical publications. One of them addressed blasphemers: they should pay a visit to Ruysch's anatomical museum and see for themselves the greatness of God's creation.[17]

On the other hand, Ruysch was a clear representative of Enlightenment thinking, with empiricism as its core method. "Only believe your own eyes" was his motto. Persistent superstitious ideas of midwives he rejected as ways to frighten women in labor; inexplicable medical cases he described exactly, concisely, and in a down-to-earth fashion. He did not speculate and hypothesize, or presuppose any supernatural causes.

As a result of his ground-breaking preparatory and technical skills, he discovered new anatomical structures. He invited those who were curious about his discoveries and preparations to his home museum so they could judge for themselves. The doors of his anatomical museum were open for anyone who wanted to see.[18]

Ruysch apparently was a pious man and, at the same time, an empirical, worldly man. He was living and working in a time that took a very mechanistic perspective on life on earth: living beings should be understood as machines "preformed" by God and set in motion, essentially created by Him from the same matter as inanimate nature and also subject to the same movements (forces).

VITAL FORCES

In the second half of the eighteenth century, the tide turned against mechanism as the dominant view of life. Natural philosophers such as Albrecht von Haller (1708–1777) concluded that a fundamental difference existed between the living and the nonliving elements of nature. The idea was that life was governed by a vital force, an "unknown principle" that was responsible for the development, growth, and regeneration of living beings. This view is called vitalism.

There were a number of reasons why people began to doubt the mechanistic conception of life, one of which was the phenomenon of regeneration. A cut in the skin, a broken bone, in some animals even an amputated leg or tail repaired itself. One scholar who was interested in this phenomenon of regeneration was the Amsterdam anatomist Andreas Bonn (1738–1817). His interest lay primarily in the regeneration (healing) of bone tissue after fracturing. For this purpose, Bonn studied and described the collection of diseased bones that the Amsterdam physician Jacob Hovius (1710–1786) had brought together since about 1750. In 1773, Hovius had donated this collection to the Surgeons' Guild, which was still located in the Snyburg. In particular, the collection would serve the surgeons' apprentices, but in order to prevent them from soiling the bones with their hands or stealing one or two, Hovius demanded that a cabinet be built in which to house them.

On the molding of this cabinet, which was filled from top to bottom with morbidly diseased bones, an adage was written: "This is Hovius's gift, which shows the healing power nature still possesses when the art of healing succumbs." The sentence is enigmatic, but understandable when one knows the anatomists' interests at that time: the healing power of nature is the regenerative capacity of the bone tissue. After Hovius's death in 1781, his portrait was added to the cabinet. Since then, Hovius has been watching over his cabinet of bones with a stern eye. Hovius's cabinet is still one of the most popular exhibits in the current Museum Vrolik.

ROMANTIC *NATURPHILOSOPHIE*

Around 1800, a historical shift occurred in the way in which vitalism gained meaning. Within the literature, art, and music of various European countries, romanticism arose as a countermovement to the rationalism of the Enlightenment. Instead of human reason and the sober mind, the romantics focused on the emotions and intuition, on creativity and fantasy. In Germany, a scientific movement sprang from romanticism: the romantic *Naturphilosophie*. Its practitioners believed that emotion and intuition should lead in the study of nature. Nature, they argued, was a unity. Stones, plants, animals, and people were an increasingly complex manifestation of a fundamental unity of mind and matter. This unity was called the *Weltseele* (world soul), the Absolute, an entity that could generally be equated with God.

Because the human mind was the highest manifestation of the *Weltseele*, man was intuitively able to understand all its other forms of expression. Man could recreate nature in his mind through the path of intuition, and thereby unravel the laws of nature. The observations made in nature, for example in anatomical research, could then be matched to these intuitively conceived natural laws.[19] It is not surprising that empiricism, which had been the leading philosophy since the days of Ruysch, was challenged by this romantic philosophy of nature, and that creative, imaginative, but also speculative theories could flourish.

VORMKRACHT

Yet not all anatomists and other naturalists in the first half of the nineteenth century—either in Germany or elsewhere—were so deeply seized by romanticism. Certainly not in the Netherlands. It is sometimes said that "true" romanticism bypassed the Netherlands because of its sober and down-to-earth Protestant Calvinist spirit. Although it is probably not that easy to generalize, it is certainly true that there were no romantic *Naturphilosophen* among the Dutch anatomists of that period. Most of them felt much more affinity with less speculative naturalists from Germany and France, such as Georges Cuvier (1769–1832), Johann Friedrich Blumenbach (1752–1840), and Karl Ernst von Baer (1792–1876). The Amsterdam anatomists Gerard Vrolik (1775–1859) and his son Willem (1801–1863), who together would dominate the Snyburg between 1800 and 1863 and for whom the Museum Vrolik was named, also shared this more sober view.

Like Blumenbach and Von Baer, the Vroliks did share elements of the romantic approach to science, such as the conviction of a teleological (goal-driven) hierarchical arrangement of life, a unity or harmony in nature, and the existence of a vital force, responsible for growth, reproduction, development, and recovery of living organisms. But they did not assert that with their own mind, based on intuition and fantasy, they could unravel the laws of nature or God's plan. The human mind was too limited for this. The only thing it could do was study the results, the consequences, of the laws of nature.[20] In that regard they shared with Ruysch a certain sense of empiricism.

In Willem Vrolik's view, living nature was driven by a life force he called *vormkracht*, a term that was analogous to Blumenbach's *Bildungstrieb*. Vrolik particularly discusses this force in his studies of teratology, the science of monsters. *Vormkracht* was an incorporeal principle that could not be observed directly or investigated fundamentally. Only its effects lent themselves to study. Through careful study of deformed humans and animals, which were seen as the results of malfunctioning *vormkracht*, Vrolik would surely also learn more about its normal workings.[21] However, one thing was crystal clear to him: the study of these effects, by means of empirical anatomical research, would also bring him a little closer to understanding God's plan.[22] Willem Vrolik writes: "In the contemplation of Nature, and in the appreciation of Her phenomena, I discern a certain plan, the consummation whereof is the preservation of the Universe. It is my conviction that acknowledgement of this plan must stand in the foreground of the practice of Science. By gazing in that direction, the persuasion is enlivened in us that an incorporeal principle, a Great Power, although unseen and usually uncomprehended, reveals itself by its industriousness as a ministering, watchful and sovereign Godhead."[23]

Willem Vrolik, the anatomist of monsters, kept teaching anatomy and doing dissections in the Snyburg until the year before he died. But during this time the inside of the space had completely changed its appearance: the theater itself had been removed—only the ceiling remained untouched. Of the collection that had been amassed since Ruysch's time there, only the cabinet of Hovius was of significance. Vrolik complained about the shortage of anatomical collections for teaching purposes and added new preparations, skeletons, and skulls. Most such objects, however, he would add to his and his father's personal collection, the Museum Vrolikianum, a collection comprising over 5,000 pieces which was housed in his father's double canal house.

Thus, just as Ruysch had done in his time, father and son Vrolik created their own private anatomy museum. But the scope, the philosophical concepts, and also the visiting policy had changed. No longer was their anatomical museum something open to all. Nor was this the case at this time for the many other smaller or larger anatomical collections, whether private or within university walls.[24] Finally, the preparation style had changed. In the first half of the nineteenth century, the injection technique perfected by Ruysch was still used—the Vrolik collection still contains hearts and other tissue injected with red or blue wax, dried or in liquor. But none of these preparations were ornamented in the ways Ruysch had done.

What the nineteenth-century anatomists like the Vroliks still shared with their predecessor Ruysch was the presence of God. His function, however, had changed: God was no longer a prominent actor within the dissection or construction of preparations, out of conviction or necessity. He now operated more in the background. He had set it all in motion, arranged the order, set the rules, and created the harmony. Men, scientists, anatomists, and the like could then study these laws and this order as if God had never been present.

But perspectives would change again. In the 1840s, a group of reactionary physiologists stood up: the more or less romantic science of the four preceding decades was a horror to them. Goal-driven vital forces were unnecessary to understand nature, they argued. They looked at physics and chemistry which had become ever more important and proclaimed that the same forces that govern lifeless nature also work in living nature. Life was a complex physical-chemical process: As Dutch physiologist Jacob Moleschott (1822–1893) said, "Ohne phosphor kein gedanke!" (No thoughts without phosphorus!) Speculative theories were forced out of science—empiricism and especially the experiment were leading. Because these physicalist physiologists reduced living nature to the same matter and force as inanimate nature, they were called reductionists. To some degree, Ruysch's Enlightenment *ratio* was back, but the mechanism had changed from dualism to monism: earth's matter and forces (living or nonliving) did not need a god. Yet it was not only physicalist physiologists who doubted God: with the emphasis on blind selection in his evolutionary theory (1859), the influential Charles Darwin (1809–1882) was also ready to give up divine teleology. But he was no radical: he hesitated to go so far as to call himself an atheist.

In 1858 Willem Vrolik, who had taught physiology in Amsterdam up to that moment, gave notice that he had no feeling for the "new" physiology. It was taken over by another scholar, the first professor of physiology of Amsterdam. Five years later Willem Vrolik died. Another five years later the anatomical theater, including all its collections, would move out of the Waag building. The ties between Ruysch and Amsterdam anatomy, embodied by the Snyburg, were thus severed in 1868.

REENCHANTED ANATOMY

The developments in physiology were not isolated. In a broader sense, increased knowledge in physics, chemistry, and medicine had resulted in the positivistic conviction that science and technology would solve all our problems, that they would be the only beneficial way for the progress of humanity. "God is dead" proclaimed Nietzsche in 1882—science was the new god. The last decades of the nineteenth century were seen as the age of science. The Nobel prizes were erected in physics, chemistry, and physiology; sciences specialized, and new disciplines and subdisciplines branched off; new laboratories were founded. Anatomy was no exception. When, in 1868, the anatomical theater left the Waag, the new institution was given the name of anatomical *laboratory*. The name change embodied the change in scientific workplace. New branches of anatomy included functional anatomy, experimental embryology, and neuroanatomy.

But at the same time as this belief in progress through science dominated the *fin de siècle*, fears also arose. Didn't all these new developments move too fast? The sociologist Max Weber (1864–1920) gave the answer: science had disenchanted the world.

Some scientists, especially biologists but also anatomists and psychologists, wanted the enchantment to return not only to the world but particularly to science: a *re*-enchanted science.[25] Some of them attempted to find a synthesis between the different scientific disciplines and specialisms, and also between science and religion or spirituality.[26]

The idea of synthesis was one of the core elements of holistic science at the beginning of the twentieth century: the biologists and anatomists were convinced that a living being (the "whole") was more than just the sum of its various parts. It is not surprising that most of these holists were developmental biologists, anatomists, psychologists, and brain scientists: after all, they investigated complex processes that were difficult to explain with the reductionist scientific method. Moreover, holistic biologists did not believe that there was only one way to come to the truth; instead, there were numerous methods, and they could come to several truths which could coexist.

In Amsterdam, holism in biology would find its most powerful champion within the walls of the newly built anatomical laboratory. The name of this holist anatomist was Louis Bolk (1866–1930). Already as a student, Bolk criticized empiricism as a science without philosophy: "nowadays many deliver the materials for construction [of the building], but few are the architects."[27] However, in the last three decades of his professorship (1898–1930) he would really emerge as a neoromantic holist.

From the 1880s onward, the Amsterdam anatomists' scientific work consisted of research in the field of evolutionary morphology, which meant that they dissected human bodies as well as those of animals—in particular monkeys and apes—and that they theorized about their descent and evolutionary relationships. It also meant that they would collect preparations and specimens in line with this research. Bolk alone amassed over 800 skulls of different monkeys and apes. But Bolk would go further than just proving evolution by dissecting and collecting. He would take evolutionary theory as a starting point for all kinds of new creative, speculative hypotheses. His best-known theory was undoubtedly his theory of fetalization. Bolk had noticed that in many characteristics (shape of skull, lack of body hair, etc.) our bodies looked much more similar to young apes than to adult apes. Bolk concluded that during human evolution, these fetal or juvenile traits were somehow conserved into adulthood. "Man may be considered as a primate fetus that has become sexually mature."[28]

Bolk took evolution as an expression of the organism itself. It was an active, internally driven process, unlike Darwinism's notion of our development as a (passive) response to external factors. Thus he understood evolution as principally a determined process, a progressive development that was an expression of a fundamental property of life. The lowest primordial organism already contained the necessity of anthropogenesis, or the process of becoming human. The entire complex of animal life on earth was one organic whole for Bolk. It had started with a basic form and then developed along fixed paths into more highly organized forms. Ultimately, anthropogenesis was reached. Bolk called his view panbiotism—philosophically, it simply satisfied his personal sense of causality. He did not want to go any further and did not impose his worldview on others: For "every form that life adopts bears the stamp of individuality."[29]

We are used to studying life through magnifying glasses and thereby bringing the otherwise invisible material within our field of vision. How completely

> *different … would our view of life be, if it were possible to us to study it with "diminishing" glasses, through which we could bring within our perspective all that is beyond what is surveyable to the naked eye. Instead of the material connections that we aim to study now, we could then much more focus on the coherence of the phenomena.*[30]

CONCLUSION

With his speculative theories and panbiotism, the early twentieth-century Bolk could not be further from Ruysch's seventeenth-century motto "believe only your own eyes." The contrast was between empiricism and neoromanticism—the desire to zoom in and discover the materials from which the body was constructed, or to zoom out and see the larger whole. In that respect, father and son Vrolik took a middle position between these extremes.

Religion remained important throughout the two centuries between Ruysch and Bolk, although Bolk's panbiotism was much more spiritually oriented than Ruysch's Calvinism, more reminiscent of Rudolf Steiner's (1861–1925) anthroposophy.[31]

Between Ruysch and Bolk, the views on science and nature bounced back and forth between rationality and emotion. In the second half of the twentieth century, they shifted again: we live once again within a scientific worldview of materialism or physicalism, where empiricism and experimentation are central.

Ruysch's preparations seem to come straight from another world, one far removed from ours, because of their aesthetics, their poetic aspects, and their attempt to defy the border between life and death. In that sense they seem enchanted; they enchant us to this day. But if we look at Ruysch's underlying motivations, his method and motto, they actually fit well with our current view of modern science. The way he promoted his knowledge makes him modern—scientists nowadays know only too well the words "valorization" and "societal impact." And there is also the business side of Ruysch. He sold his collection just as scientists today try to capitalize on their scientific results through the foundation of small companies, or by having their research funded by the business community.

Finally, what makes Ruysch modern is the fact that his museum was open to visitors. As I have mentioned, in the nineteenth century anatomical museums closed their doors to the general public. This remained the situation up till the last decades of the twentieth century. Only then did some of these museums (re)open their doors for everyone to see, as in Ruysch's day.

Even today, however, many of the anatomical and pathological museums in the world are inaccessible to nonmedical people, and for various reasons. I think that is a shame. Anatomy museums are museums of the body and of the way we have come to know it in history. Since we all have a body, anatomical museums offer a unique, historically contextualized way to learn about that body, about the way it functions, about disease and health and about matters of life and death. As such, these museums allow us to contemplate what it means—and has meant—to be human.

Frederik Ruysch: Philosopher of Divine Decay

STEPHEN ASMA

GIVEN HIS DATES (1638–1731) AND HIS MEDICAL ACCOMPLISHMENTS, IT is common to classify Frederik Ruysch as an Enlightenment hero of natural science. He was almost a perfect contemporary of the long-lived Isaac Newton (1642–1727), of whom Alexander Pope quipped "God said 'Let Newton be!' and All was Light." In fact, Ruysch lived so long that he bridged the transition between premodern supernaturalism and modernism's mechanization of nature.

In what follows I want to suggest that Ruysch had one foot squarely in the older world of hermetic occult knowledge, or arcana, even as he helped usher in exoteric medical knowledge. Ruysch did not singlehandedly change the paradigm, but we can see in his work the culture-wide shifts in epistemology (i.e., what counts as knowledge), social epistemology (i.e., who has knowledge), and metaphysics (i.e., what counts as causation). These seemingly objective and instrumental uses of nature are not independent of values. Even the mechanistic empiricism of the eighteenth century is tinged with moral normativity: Is there design in nature? Why is life so ephemeral? Why is there so much suffering in biological processes?

In addition to arguing that Ruysch is a bridge between the old enchanted and modern disenchanted views of nature, I also want to suggest that his work and collections reveal a kind of modern "immanence philosophy." Immanence philosophy, from Aristotle through Spinoza, argues that God is not just the Creator who lives above and beyond Nature (i.e., a transcendent view of the divine), but instead God or the spiritual is indwelling or inherent inside Nature (i.e., immanent). Ruysch's medical practice and aesthetic creativity evince a uniquely modern immanence philosophy, and he may be one of the most sublime fusions of material and spiritual views of nature. Ruysch's dioramas and specimens look for beauty and even redemption amidst the grotesque, the sorrowful, the decaying, and the tragic. The unique immanent philosophy of Ruysch is brighter and more optimistic than medieval fatalism, but it is also darker than the naively optimistic Enlightenment philosophy of his late contemporaries.

To understand Ruysch's place in the transition from hermetic premodernism to the mechanistic worldview, we need to understand the changing relationship between knowledge, God, and nature. When Ruysch was young it

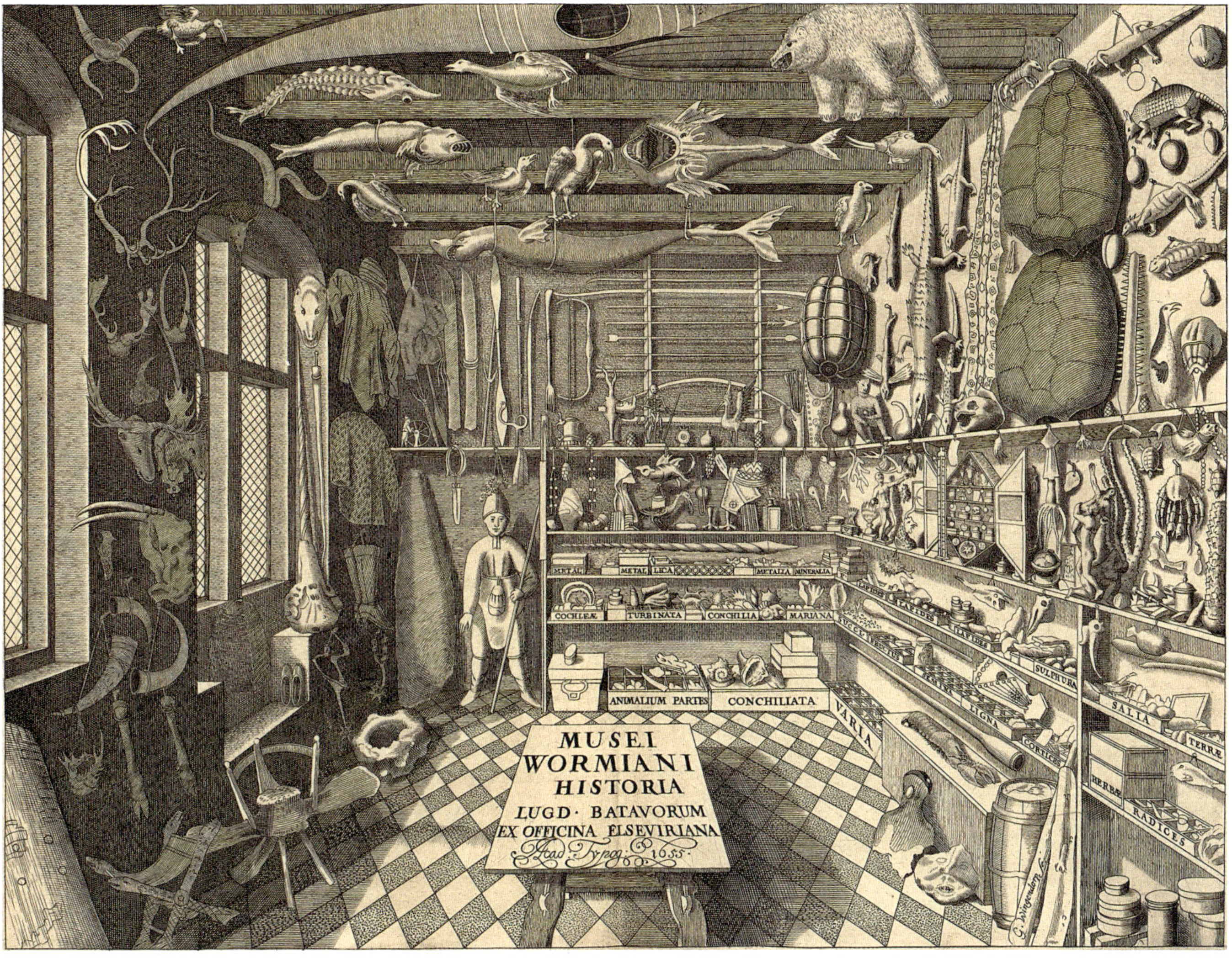

Image above: Ole Worm's curiosity cabinet, from the frontispiece of Museum Wormianum, *1655.*

was still common for natural philosophers to see nature through the lens of the *Emerald Tablet*, a foundational alchemical text that purported to date back to the legendary Hermes Trismegistus, a Hellenistic hybrid character, part Hermes and part Egyptian God Thoth. During the Hellenistic period (circa 323–31 BCE) the Hermetic canon (papyri texts) grew, focusing on magical incantations, totemic manipulation, and the animation of statues and icons. Neoplatonism and Gnosticism were influenced by this magical tradition, but there was a powerful Hermetic revival in the Renaissance period (notably in Marsilio Ficino's translation of the Corpus Hermeticum) and again in the early modern period (notably in Thomas Browne's *Pseudodoxia Epidemica*, 1646–1672).

The core idea at the center of this spiraling system of hermetic texts and practices is summed up in the first few lines of the legendary *Emerald Tablet*:

"Quod est inferius est sicut quod est superius, et quod est superius est sicut quod est inferius, ad perpetranda miracula rei unius" (Whatever is below is similar to that which is above, and that which is above is like that below, to work the wonders of one thing).[32] And the "one thing" referred to here is indeed the One—the great unifying unconditioned. It is the Parmenidean One and the monotheistic God rolled together. The first true metaphysician Parmenides (c. 515–? BCE) argued that beneath the fluctuating world of appearance (wherein human senses and opinions flourished) existed the changeless reality of the One. This ultimate and hidden One or Being formed the conceptual foundation for all the changes we perceive in nature. In later monotheism the One was retranslated into Yahweh or Allah, but in the hermetic tradition it retained its nonsectarian and mystical obscurity.

Ruysch, Newton, and other supposedly strict empiricists were still under the sway of this theological view of nature.[33] God or the One creates the cosmos by a force of emanation, like a conflagration lighting up and heating up a whole region. The One is the source of being and radiates out degrees of reality, creating nature and everything in it. According to this magical view, chains of correspondence exist between the heavenly realities and the objects down here. There is a great chain of being, but also an isomorphism—or formal correspondence—of macrocosm and microcosm forms. As above, so below. There are correspondences, for example, between natural creatures, parts of the anatomical body, and astrological patterns (e.g., the scorpion creature and the constellation Scorpius were mapped onto the pubic or libidinal area of the human body, just as the lion and Leo constellation are said to have correspondence in the human heart/chest region, and so on throughout all of nature).

Once the magi or virtuosi understood the signs and the correspondences, then mundane objects of nature could become talismans for reaching upward (i.e., selfless or high magic, sometimes called white magic) or they could be used for manipulation of the mundane world (i.e., low or black magic). If you were using the signs and symbols in nature piously, you were healing others and/or communing indirectly with the transcendent One. An early form of isomorphic healing can be seen in the Hebrew scripture Numbers 21:4–9, where Yahweh instructs Moses to create a talisman—a staff with a snake sculpture on it—in order to heal the snake bites suffered by Moses's people. An only slightly modified logic was still viable in medicine in Paracelsus's sixteenth-century pronouncement that *similia similibus curantur* (what makes a person ill, also cures them), and even the eventual homeopathy movement of the late eighteenth century, which continues today, assumes the much older idea that "like cures like."

Ruysch's generation slowly and imperfectly transformed this older view of symbolic nature and immanent God into a mechanical nature and deistic God.

Solutions for nature's mistakes and malfunctions were not to be found through bleeding patients at points of astrological-anatomical convergence, or giving talismanic tinctures; rather the solutions would need to come from inside the mechanical system itself (i.e., the self-contained and self-justifying nature). Pathologies, teratologies, and diseases needed to be remedied by working horizontally across mundane cause and effect (i.e., surgeries and physiological interventions), not vertically up to a benevolent metaphysical One.

Ruysch first trained as an apothecary in The Hague, eventually buying his own shop in 1660. He used profits from his shop to study medicine in Leiden, where he developed some of his early anatomical and specimen preparation skills, earning his M.D. in 1664. Ruysch moved to Amsterdam to serve as a governing member of the Surgeons' Guild, where he continued to acquire new skills—this time studying and mastering obstetrics. This expertise brought him into contact with midwives, whom he taught relevant anatomy. As a medical practitioner, Ruysch was respected and renowned, although he became embroiled in the usual professional battles—including a very public debate with another obstetrician in 1677 regarding culpability in the case of a patient's torn perineum. His immersion in the world of obstetrics and midwifery (he was even appointed the official city obstetrician) undoubtedly gave him unique access to unviable fetuses, miscarriages, and even infanticide cadavers—all of which eventually populated his various tableaux.

All the while Ruysch was mastering and practicing medical skills, and acquiring botanical acumen as well, he was also building his famous anatomical collection or cabinet. The cabinet was housed in several rooms of Ruysch's home at 15 Bloemgracht Canal. Curiosity cabinets were well known, of course; recall that Danish physician Ole (Olaus) Worm (1588–1654) catalogued his collection in the 1650s in *Museum Wormianum*.

At the same time, Rembrandt had a substantial curiosity cabinet in his home studio—a short walk from Ruysch's home in Amsterdam—to elicit wonder of course, but also to help the artist render exotic accoutrement. Another contemporary, Galileo Galilei (1564–1642), was notably irritated by the collecting mania bursting out everywhere with the age of exploration. According to Galileo, too much obsession with wonderful variety obscured the underlying laws of nature. He had disdain for "some little man with a taste for curios who has been pleased to fit it out with the things that have something strange about them because of age or rarity or for some other reason, but are, as a matter of fact, nothing but bric-a-brac."[34] A proper collection, according to that great champion of heliocentrism and efficient causality, should have a purpose. And Ruysch did have a purpose, indeed a couple of purposes, for his cabinet.

Medical knowledge, especially medical anatomy, was best taught and understood through visual means, and Ruysch's cabinet was the visual method.

Even today the visual arts are integral to medical anatomy. Ruysch's specimens could greatly improve medical understanding and practice, in part because they were extremely lifelike and gave the viewer a realism and physiological insight hitherto only seen in vivisection. Ruysch patiently injected liquified wax into the delicate organic structures he uncovered during dissection. He found a particularly subtle recipe of wax, talc, and cinnabar (mercury sulfide). And he developed a long-lasting embalming fluid, *liquor balsamicus*, for wet preparations. Carefully guarded by Ruysch, the embalming liquor was probably a pepper-infused distillation of wine or corn (or possibly Nantes brandy). Historically, these harsh alcohol baths (spirits) distorted and damaged specimens, but Ruysch found a gentler, salubrious blend. A generation later, John Hunter, the Scottish anatomist living in London, finally rivaled Ruysch's skill in preservation with his own secret spirits and injections—still on display at the Royal College of Surgeons in Lincoln's Inn Fields.

Visitors to Ruysch's cabinet were common enough, but 1697 brought Peter the Great of Russia, who was so impressed by the quality of preservation that he returned twenty years later (after lengthy statecraft distractions) and purchased the entire cabinet for around 30,000 guilders, a huge amount at the time,[35] shipping it back to Russia along with Albertus Seba's equally impressive Amsterdam cabinet. Ruysch's secret distillation and preservative recipe was supposed to be sold to Peter with the collection, but no credible copy has ever been located.[36]

Peter the Great celebrated teratology—the study of bodily abnormalities—in his early museum collections, and in some of his official proclamations prohibited the killing of deformed children, requesting that local officials send the "marvels" to his St. Petersburg museum. Peter was himself a marvel, at 6 feet 7 inches, and may have purchased a French giant named Bourgeois to become his constant companion (his personal servant) because Bourgeois made Peter look "normal" by comparison.[37] Over 900 of Ruysch's specimens still remain at the Peter the Great Museum of Anthropology and Ethnography (Kunstkamera) in St. Petersburg. Many of these have been photographed and can be viewed online.[38]

Ruysch's preservation of cadavers, body parts, and the fine elegant structures of vascular systems was not simply the mechanical ingenuity of a lab technician. It was literally and figuratively secret knowledge—gnostic virtuosity. Ruysch was part of a long hermetic tradition that slowly became the more recognizable scientific project, namely making the invisible visible. For example, he literally made the obscure valves in the lymphatic system visible, via his preparation of specimens, and thus resolved a crucial debate of the day.[39] But he was also very secretive about some of his innovations. Legend has it that he sometimes blew out the candles during a delivery, midwifing a baby in the dark lest competitors see his innovative obstetric tools.

In what way, however, is Ruysch an immanence philosopher, finding divinity in decay? He did not write philosophy per se, and he generally loathed theory. But he expressed himself and his *Umwelt* through his powerful exhibits. The tableaux, famously rendered by Cornelius Huyberts for the *Thesaurus anatomicus* and the *Opera omnia*, are well-known examples of Ruysch's blend of *vanitas* tradition and scientific realism. Fetal skeletons playing music in a dense ecology of coiled intestines, dried blood vessels, and bladders, or crying into a clothlike mesentery, or holding a mayfly or string of pearls—these demand attention and contemplation. These dioramas were not mute indecipherable structures, since Ruysch included memento mori Latin phrases, like this one from the book of Job: "Homo natus de muliere, brevi vivens tempore, repletur multis miseriis" (Mortals born of a woman live only a short time and are full of misery). Or another text: "Cur ea diligere velim, quae sunt in mundo?" (Why should I wish to regard the things of this world?).

Sadly, none of these dioramas survived, but consider the many beautiful wet-prep jars that still reside in St. Petersburg. Here one finds the severed hand of a child dressed elegantly in a lace cuff (possibly sewn by Ruysch's daughter), and this severed hand is gently holding a human fetus. It is hauntingly beautiful, bearing no sign of the violence or craft that gave life to it.

These remarkable displays are revelations, but of what? They are hybrids of two worlds—the hermetic and the factual—and once we understand this, we can grasp the immanence philosophy of Ruysch. First, we must note that specimens are exemplars, types, samples, or tokens. A "type specimen" may be privileged in the sense that a new species description and name are based upon a particular specimen, but even then the cadaver or object could be hypothetically replaced by another from the same family and it would still function properly as a representation of a whole population of similar creatures. Many objects in a collection have this same function, whether it's a "unicorn horn" in a Renaissance *Wunderkammer* or "Sue" the *T. rex* in Chicago's Field Museum. Contrast this with a relic, like a bone of St. Peter, or a piece of the true cross. These objects of curiosity are prized in a collection, but they are not "types" at all. They are not kinds of things, but radical individuals. The specimen and the relic are both curios, but quite different as signifying objects.

The artwork object is closer to the relic, in a sense. It is one of a kind. It may point beyond itself to something else—a kind of beauty, a kind of injustice, a kind of technique—but it is not fungible. Amazingly, Ruysch's carefully crafted exhibits are all of these things. The wet-preparation lace-cuffed hand with a fetus is an utterly unique artwork, an object of emotional veneration, and a specimen revealing real medical data. In this, Ruysch follows the remarkable blending of art and science first revolutionized by Andreas Vesalius (1514–1564).

The cherished collectable, whatever it is, always points beyond itself. In a premodern paradigm the object points to hidden spiritual truths and systems of esoteric meaning. In the modern paradigm, it moves from the specimen object to the hidden law, cause, or principle of nature. A mayfly (*Rhithrogena germanica*) in a modern post-Linnaean collection points to its phylogenetic placement as a member of the phylum Arthropoda, the class Insecta, and the order Ephemeroptera. But in a premodern display of a mayfly—like the one held in the hand of the posed fetal skeleton in Ruysch's diorama—the dried insect points to or symbolizes the brevity of all life. It becomes a memento mori. The mayfly, which only lives about twenty-four hours, is a symbol or microcosm of the macrocosm. This is the hermetic specimen or exhibit. It is not an indecipherable sign, like the encrypted codes of secret societies (e.g., Rosicrucians, Freemasons, etc.), but a simpler step from appearance to reality. Ruysch, like his friend Seba who included the biblical hydra in his natural history, is still connected to the iconographic view of nature—nature as moral message. Like all hermetic thinkers, Ruysch agrees that everything is really something else.

In the hermetic tradition of arcana, intuitive insight (synthetic and creative) is prized over discursive deduction (analytic and destructive). Imagination is not simply fantasy, but actually the creative force of knowledge, revealing our divine aspect—the alchemists only typify this more widely held notion that creation through transformation (i.e., distillation and mixing) is godlike and within reach for cognoscenti virtuosi. Ruysch himself, with his ability to reveal secret organic tissues and make flesh eternal, is a more sober and serious alchemist. The influence of prescientific thinking is further evidenced by the way embryologists, including Ruysch, take seriously the idea that a pregnant mother's imagination can cause "monstrous" or teratological distortions in the developing fetus.

Moreover, medical anatomy has the same commitment—not shared for example by physics—that pictures or images are epistemic ladders to reach upward from the senses to the intellect. The power of imagery, long discounted as dangerous by the aniconism in many Western religions, was kept alive by the hermetic traditions. Images were dealt an additional blow by the rise of mathematical physics, but anatomy and other descriptive sciences could not dismiss imagery. Instead, anatomists like Ruysch doubled down on the power and knowledge of carefully wrought images and didactic objects, placing their new empirical observations inside the older allegorical pictorial tradition. Witness for example the way Cornelius Huyberts's frontispiece for Ruysch's *Opera omnia* (see page 104) is loaded with symbolic icons: Father Time, personified Dame Nature, olive branches, cornucopia, and so on. This ties Ruysch's work explicitly to the premodern view of a moral and symbolic nature, typified and codified in Cesare Ripa's influential *Iconologia* (1593, 1603).

However, even as Ruysch moralized nature and reveled in the *vanitas* traditions of the premodern paradigm, he also slowly ushered in the disenchanted world of modernism. He was devoted to intense observation, and his writings eschew theoretical speculations. His goal was to understand the body better, ameliorate suffering, and describe a relatively mechanical biology of development. Even though he saw moral messages in nature—like memento mori—he did not think that deformed babies and illnesses were punishments from God. His own daughters—great painters Anna and Rachel—seem to complete the transition from premodern to modern paradigm, by leaving out the usual symbols of *vanitas* painting (e.g., skulls, extinguished candles, hourglasses, etc.) and focusing instead on the flowers themselves. Whatever moral message is contained in nature, it is becoming a mere whisper in the modern representations of nature.

Ruysch is part of the cohort, including people like Robert Boyle and Robert Hooke, that moved from alchemical desires to mix and distill cosmic opposites (for making gold and homunculi) to just mixing and distilling real natural substances for mundane medical functions. The magical alchemists were breaking down the material world to its *rationes seminales*, but now Ruysch and his contemporaries were breaking it down to its microscopic layers, for the purpose of medical advancement and not cosmic aggrandizement.

The hermetic view held that everything was inside everything else. The magi and the alchemists (and even the witches) did not create *de novo* (new) realities (since only God can create *ex nihilo*—out of nothing), but rather they found the seeds of one thing inside another thing and activated those dormant seeds—causing the transformation of a base metal into a rare element, or changing a man into a beast. Ruysch and his generation tempered these supernatural projects and redefined science to be a kind of magic, and vice versa. As Tommaso Campanella (1568–1639) characterized the new fusion: "A 'work of magic' is everything done by scientists who imitate nature or help it along with some unknown technique: not only the common people but the whole human community says so, the result being that not only the sciences as usually described but also all the others are in the service of magic. … As long as technique is not understood, they call it 'magic,' but afterward it is ordinary science."[40]

Ruysch is no longer trying to connect anatomy to astrology, nor is he arrogant enough to pursue the purposes (or teleology) of the deity in nature's patterns. But the body is still a sign system, and knowing how to read the signs properly was a matter of life and death. The old hermetic semiotics of the body looked for correspondence to cosmos, but now Ruysch was taking science inside the body to a new semiotics of correspondence between signs and health. Typical samples of Ruysch's new empirical expertise—a new virtuoso approach to anatomical signs—can be found in *The Celebrated Dr. Frederic*

Ruysch's Practical Observations in Surgery and Midwifery. In *Observation XXVI*, Ruysch excoriates a fellow surgeon for misreading important anatomical signs. "The wife of a certain butcher having lately brought forth a strong infant, the uterus itself immediately followed the placenta, and was inverted," explained Ruysch. He continued, "hereupon an ignorant medicaster [quack] was called … [and] instead of replacing the uterus into its proper situation, he cut a little way into the mass" trying to assess what it was. Exasperated by this ignorant approach, Ruysch stated "This rash and unheard of method of inquiry soon put an end to the patient's life, for so profuse an hemorrhage immediately followed, as was not supportable, nor capable of being suppressed by all my endeavors with the most powerful stagnotics; but everything proving fruitless, in a few hours after I was called, faintings came on, and toward the latter end of the day, the patient expired."[41]

The immanence philosophy of Ruysch consists in an almost medieval pessimism about this earthly realm—the vale of tears—combined with a heroic optimism concerning medical innovation and a determined pursuit of aesthetic beauty amidst the macabre. This is a strange, almost paradoxical set of values. If we encountered Ruysch in the twentieth century, we'd align him with existential absurd heroes. He is like a doctor on the Titanic, healing people while the ship slowly goes down.

His dioramas remind us that life is inescapably short and miserable and we must look to the world beyond, but then his life's labor is the improvement of this earthly mess. Ruysch does not have the full Enlightenment optimism of, say, Immanuel Kant, but neither does he revel in the sad teratology of his anatomical brethren. From Ambroise Paré (1510–1590), Ulisse Aldrovandi (1522–1605), up through the Hunter brothers and Willem Vrolik (1801–1863), anatomy collectors had an understandable penchant for malformed pathology specimens. Embryological "monsters" have always made for intoxicating specimens, but also helped reveal the mechanisms and patterns of normal development. Compared to the total number of specimens, however, Ruysch did not collect many aberrations per se.[42] He was the first to successfully describe several rare anomalies, including intracranial teratoma, enchondromatosis, and Majewski syndrome, but his main concentration was not abnormality.[43]

The immanence approach looks for beauty, elegance, and grace amidst the ephemeral nature of life and the inevitability of defeat and death. The most beautiful innocent creatures, namely babies, are juxtaposed against the inevitability of death. The innocence and beauty of infants as inanimate objects heightens the poignancy and elicits a philosophical mood. Ruysch did not celebrate all the ways nature can "go wrong," so to speak—whereas we see such powerful albeit disturbing teratology collections still available at the Vrolik Museum, the Mütter Museum, or the Hunterian. Ruysch did not live in a Darwinian metaphysics, where every organic form was a lucky amalgamation of blind mechanical laws (chance variation and natural selection). Rather, Nature—disappointing though it was—still had values like justice and beauty within it, because after all it was the result of a benevolent creator. Ruysch's work seems devoted to finding that justice and beauty, hidden in the body, and elevating it—the low made high. Ruysch's contemporary Isaac Newton and many others took the same deistic view of nature, but their astronomy and physics afforded them the luxury of avoiding the gore of life. Not so for Ruysch, who performed a record-breaking thirty-one Christmastime public dissections in Amsterdam. But the lurid and ignoble potential in these events was managed and elevated by placing the dissections in an aesthetic and moral framework. Dissections were often accompanied by flute music, for example, and it was made clear in the fliers and in the anatomical theater that the bodies being disassembled were "evil doers," executed criminals.

Giacomo Leopardi's "Dialogue between Frederick Ruysch and His Mummies" (1827) grasps the strange optimism in Ruysch's macabre work, when he makes the reanimated fetuses—suddenly come to life, singing and dancing—into charming and humorous interlocutors, rather than horrifying monsters. The mummies reassure the fictionalized Ruysch that death is not horrible, and dying is like going to sleep. Leopardi is recapitulating Ruysch's own philosophy, that death is not synonymous with horror.

Sociologist Max Weber (1865–1920) famously wrote, "The fate of our times is characterized by rationalization and intellectualization and, above all, by the disenchantment of the world."[44] Against this backdrop we find contemporary anatomical displays like Gunther von Hagen's wildly popular *Body Worlds*. Von Hagen created a remarkable plastination technique in the 1970s that would have made Ruysch proud. But we find the displays of anatomized cadavers in *Body Worlds* more in the style of P. T. Barnum than of a Ruysch diorama. Bodies are juggling, while riding bicycles, or playing cards, or dancing. For Frederik Ruysch the world was not yet disenchanted, and you can see that and feel that in his remarkable work. The flesh, for Ruysch, is infused with the divine—it is not ridiculous or trivial.

At the beginning of Goethe's *Faust*, God, the angels, and Mephistopheles assess the depressing human condition. The sublime misery of earthly life is partly due to Nature's capriciousness, but mainly because humans continually torment each other and themselves. Even the light of reason—a gift from God—is used by men to act "far beastlier than any beast." God asks, do we find nothing right on earth? And Mephistopheles says that man's misery moves even himself the devil to pity, and he has "scarce the heart to plague the wretched creature."

They turn then, as in the book of Job, to the character Faust for an interesting case. But while Faust subsequently plays the magical, alchemical fool and tragic paragon—becoming his own adjective ("Faustian") through literary history—we cannot forget that God spares this wretch in the end. After foolishly and selfishly using alchemy to chase women through history and engage in hedonistic play—causing misery to all in his path—Faust eventually turns his secret knowledge toward the good of human beings (i.e., white magic). He learns that knowledge of the hidden springs of nature is power, and such power only redeems when it emancipates others. Ruysch and Faust are similar heroes, in the sense that they struggle against overwhelming misery and mystery, but they are redeemed and forgiven their prideful ambitions. In part, we celebrate them because their absurd striving casts a tiny sliver of light and progress into the dark earthly realm—this is the official hagiography. But it's the striving itself that is beautiful and redemptive. Faust's passionate pursuit has made a mess of his life, but his will to understand and to create is somehow awesome. Ruysch had a tragic perspective, but thankfully his life was not particularly tragic. Ruysch the polymath and other absurdist savants seem to be Goethe's inspiration when he leads a choir of angels to announce at the end of the play, "Whoever strives in ceaseless toil, him we may grant redemption."

Dialogue Between Frederick Ruysch & his Mummies

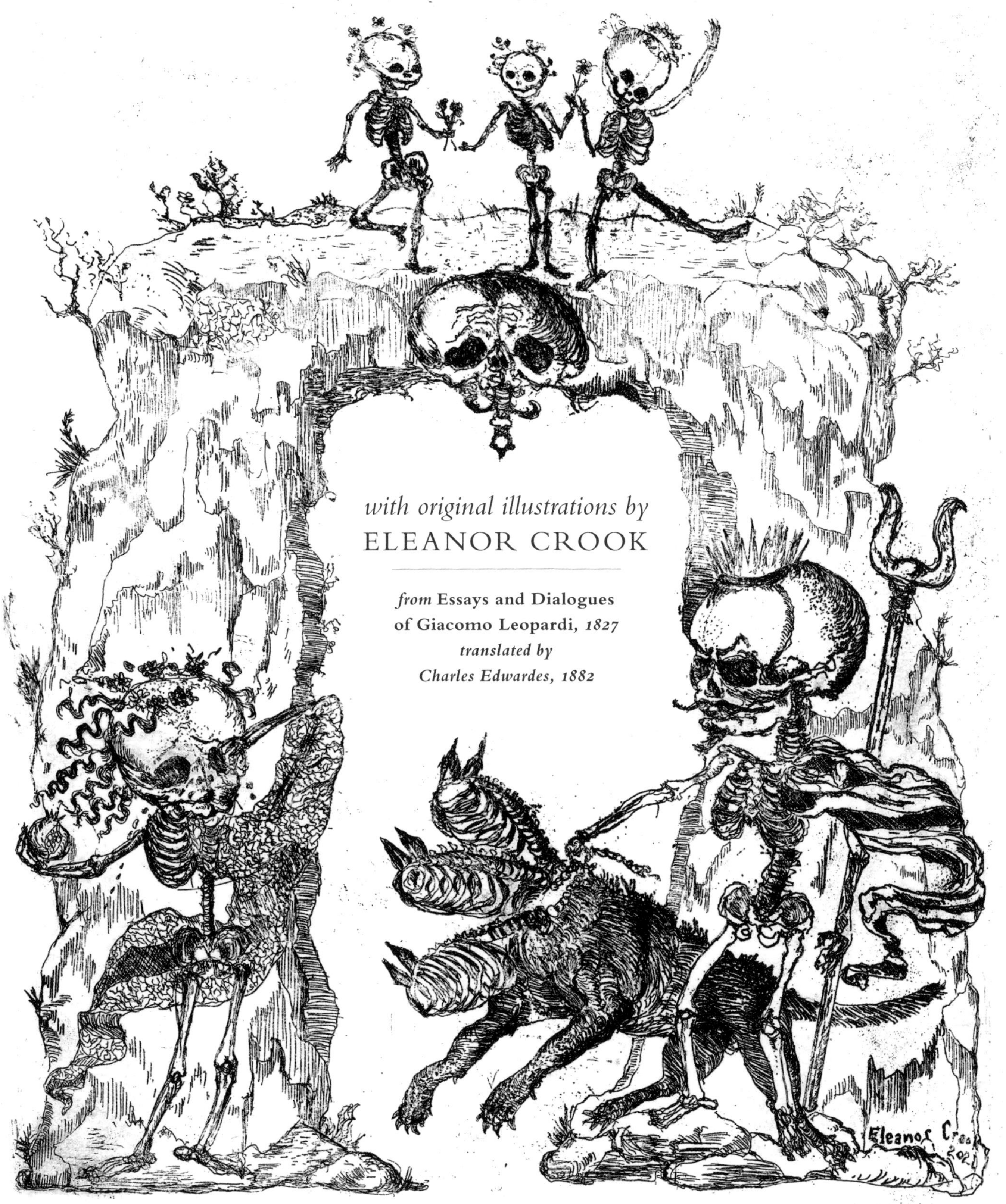

with original illustrations by
ELEANOR CROOK

from **Essays and Dialogues of Giacomo Leopardi,** *1827*
translated by
Charles Edwardes, 1882

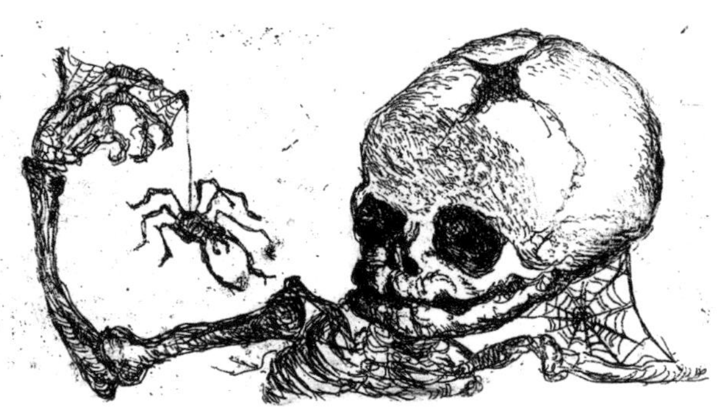

Chorus of the dead in Ruysch's laboratory.

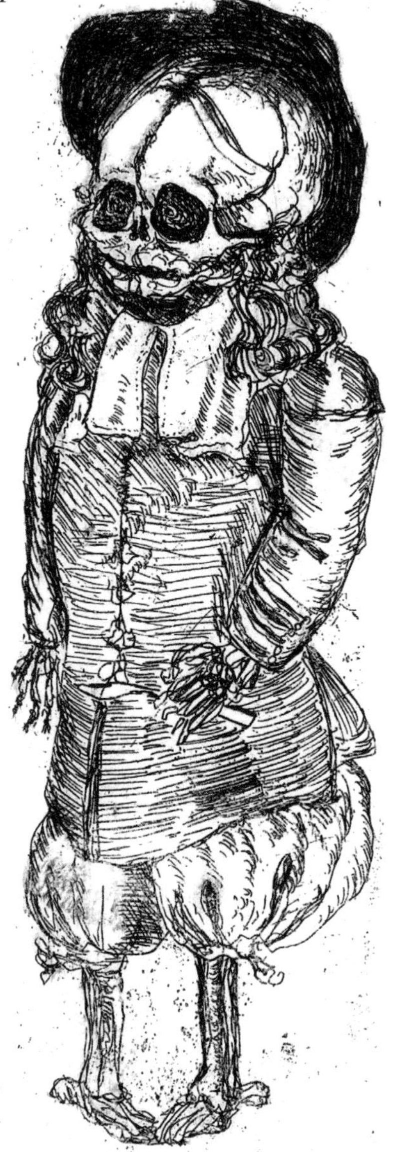

O Death, thou one eternal thing,
That takest all within thine arms,
In thee our coarser nature rests
In peace, set free from life's alarms:
Joyless and painless is our state.
Our spirits now no more are torn
By racking thought, or earthly fears;
Hope and desire are now unknown,
Nor know we aught of sorrow's tears.
Time flows in one unbroken stream,
As void of ennui as a dream.
The troubles we on earth endured
Have vanished; yet we sometimes see
Their phantom shapes, as in a mist
Of mingled thought and memory:
They now can vex our souls no more.
What is that life we lived on earth?
A mystery now it seems to be,
Profound as is the thought of death,
To wearers of mortality.
And as from death the living flee,
So from the vital flame flee we.
Our portion now is peaceful rest,
Joyless, painless. We are not blest
With happiness; that is forbid
Both to the living and the dead.

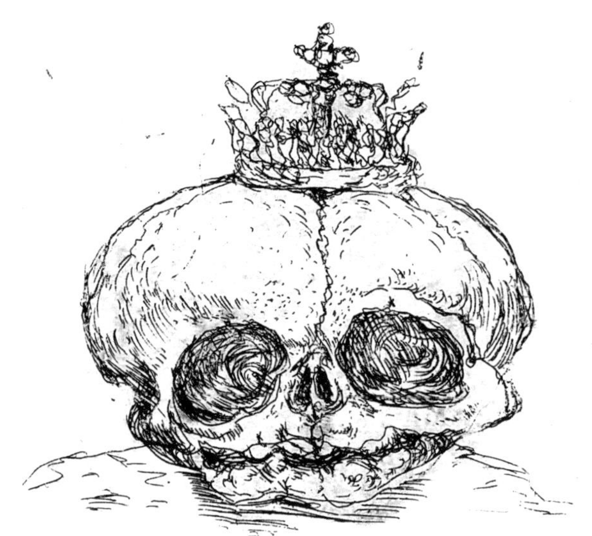

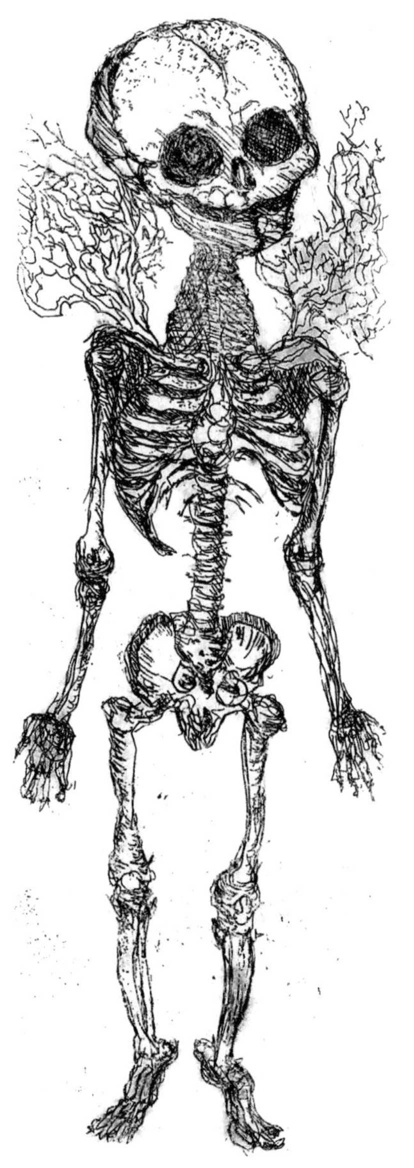

Ruysch. (outside his laboratory, looking through the keyhole). Diamine! Who has been teaching these dead folks music, that they thus sing like cocks, at midnight? Verily I am in a cold sweat, and nearly as dead as themselves. I little thought when I preserved them from decay, that they would come to life again. So it is however, and with all my philosophy I tremble from head to foot. It was an evil spirit that induced me to take these gentry in. I do not know what to do. If I leave them shut in here, they may break open the door, or pass through the keyhole, and come to me in bed. Yet I do not like to show that I am afraid of the dead by calling for help. I will be brave. Let us see if I cannot make them afraid in their turn.

(Entering.)—Children, children, what game are you playing at? Do you not remember that you are dead? What does all this uproar mean? Are you so puffed up because of the Czar's visit, that you imagine yourselves no longer subject to the laws of Nature? I am presuming this commotion is simply a piece of pleasantry on your part, and that there is nothing serious about it. If, however, you are truly resuscitated, I congratulate you, although I must tell you that I cannot afford to keep you living as well as dead, and in that case you must leave my house at once. Or if what they say about vampires be true, and you are some of them, be good enough to seek other blood

to drink, for I am not disposed to let you suck mine, with which I have already liberally filled your veins. In short, if you will continue to be quiet and silent as before, we shall get on very well together, and you shall want for nothing in my house. Otherwise, I warn you that I will take hold of this iron bar, and kill you, one and all.

A Mummy. Do not put yourself about. I promise you we will all be dead again without your killing us.

Ruysch. Then what is the meaning of this singing freak?

Mummy. A moment ago, precisely at midnight, was completed for the first time that great mathematical epoch referred to so often by the ancients. Tonight also the dead have spoken for the first time. And all the dead in every cemetery and sepulchre, in the depths of the sea, beneath the snow and the sand, under the open sky, and wherever they are to be found, have like us chanted the song you have just heard.

Ruysch. And how long will your singing or speaking last?

Mummy. The song is already finished. We are allowed to speak for a quarter of an hour. Then we are silent again until the completion of the second great year.

Ruysch. If this be true, I do not think you will disturb my sleep a second time. So talk away to your hearts' content, and I will stand here on one side, and, from curiosity, gladly listen without interrupting you.

Mummy. We can only speak in response to some living person. The dead that are not interrogated by the living, when they have finished their song, are quiet again.

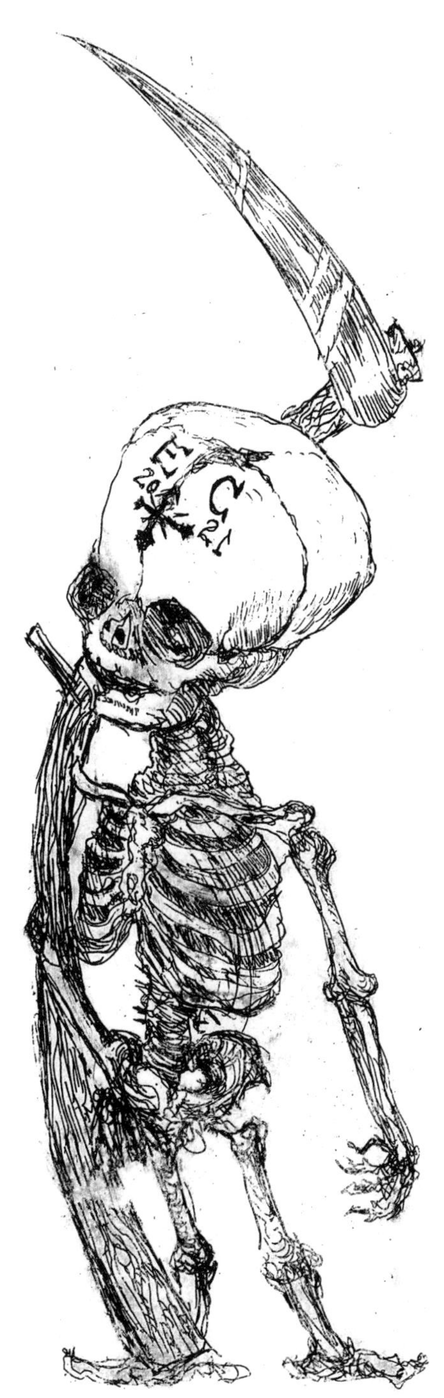

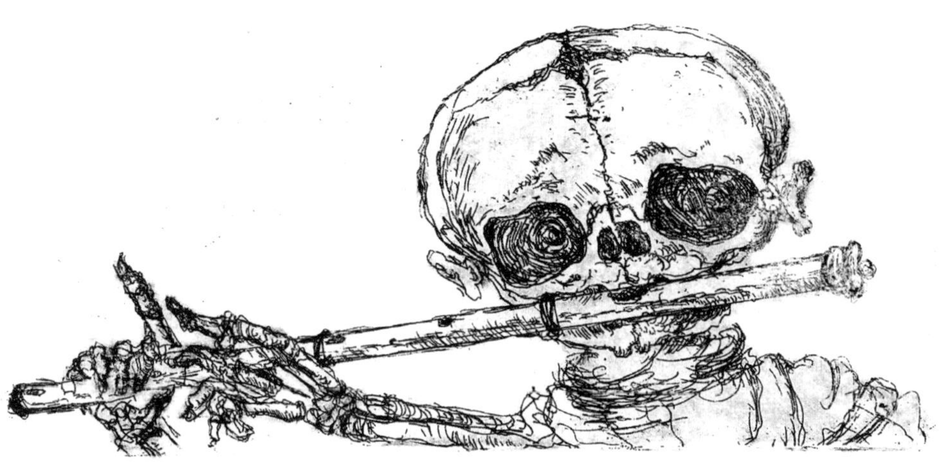

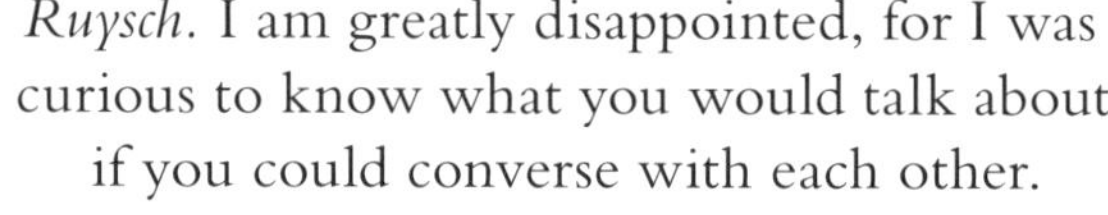

Ruysch. I am greatly disappointed, for I was curious to know what you would talk about if you could converse with each other.

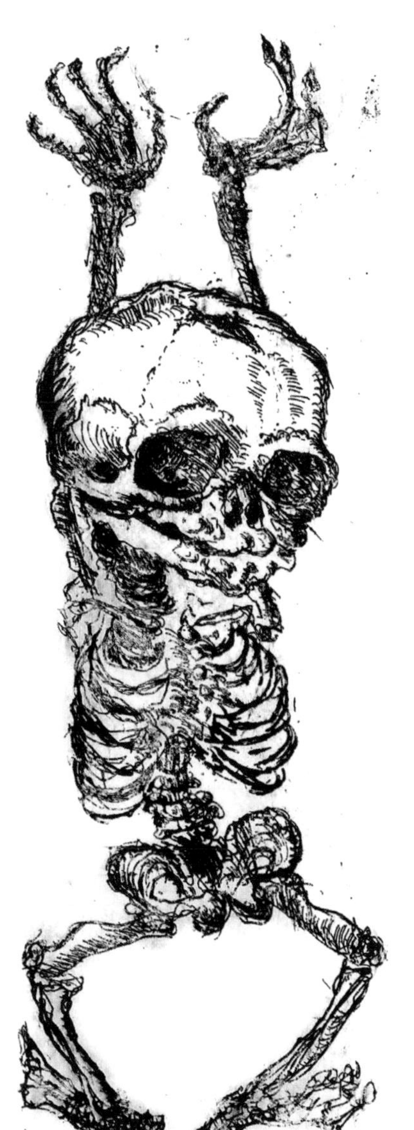

Mummy. Even if we could do so, you would hear nothing, because we should have nothing to say to one another.

Ruysch. A thousand questions to ask you come into my mind. But the time is short, so tell me briefly what feelings you experienced in body and soul when at point of death.

Mummy. I do not remember the exact moment of death.

The other Mummies. Nor do we.

Ruysch. Why not?

Mummy. For the same reason that you cannot perceive the moment when you fall asleep, however much you try to do so.

Ruysch. But sleep is a natural thing.

Mummy. And does not death seem natural to you? Show me a man, beast, or plant that shall not die.

Ruysch. I am no longer surprised that you sing and talk, if you do not remember your death.

"A fatal blow deprived him of his breath;
Still fought he on, unconscious of his death"—

as says an Italian poet. I thought that on the subject of death you fellows would at least know something more than the living. Now tell me, did you feel any pain at the point of death?

Mummy. How can there be pain at a time of unconsciousness?

Ruysch. At any rate, every one believes the moment of departure from this life to be a very painful one.

Mummy. As if death were a sensation, and not rather the contrary.

Ruysch. Most people who hold the views of the Epicureans as to the nature of the soul, as well as those who cling to the popular opinion, agree in supposing that death is essentially a pain of the most acute kind.

Mummy. Well, you shall put the question to either of them from us. If man be unaware of the exact point of time when his vital functions are suspended in more or less degree by sleep, lethargy, syncope, or any other cause, why should he perceive the moment when these same functions cease entirely; and not merely for a time, but forever? Besides, how could there be an acute sensation at the time of death? Is death itself a sensation? When the faculty of sense is not only weakened and restricted, but so minimized that it may be termed nonexistent, how could anyone experience a lively sensation? Perhaps you think this very extinction of sensibility ought also to be an acute sensation? But it is not so. For you may notice that even sick people who die of very painful diseases compose themselves shortly before death, and rest in tranquillity; they are too enfeebled to suffer, and lose all sense of pain before they die.

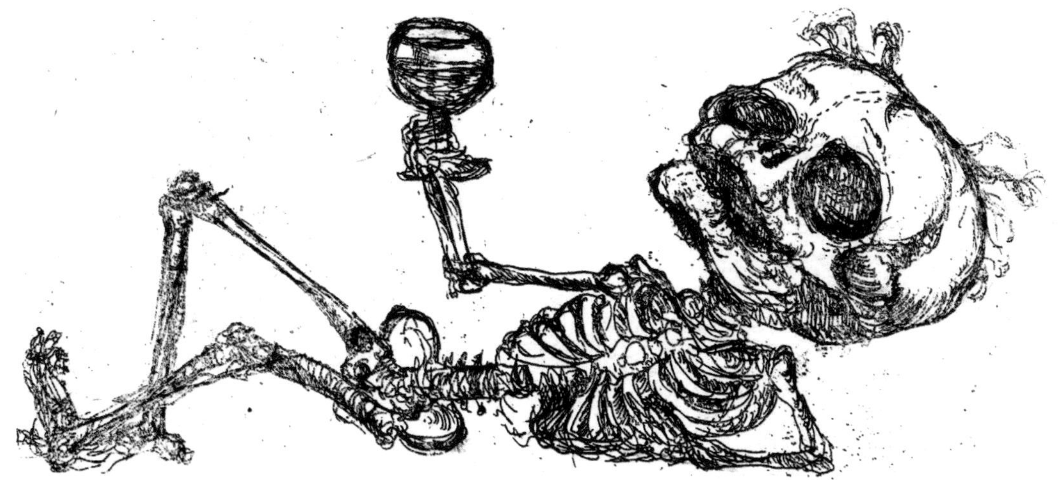

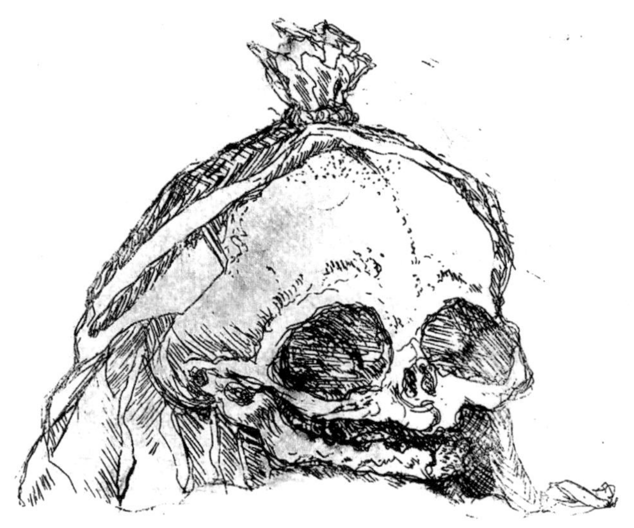

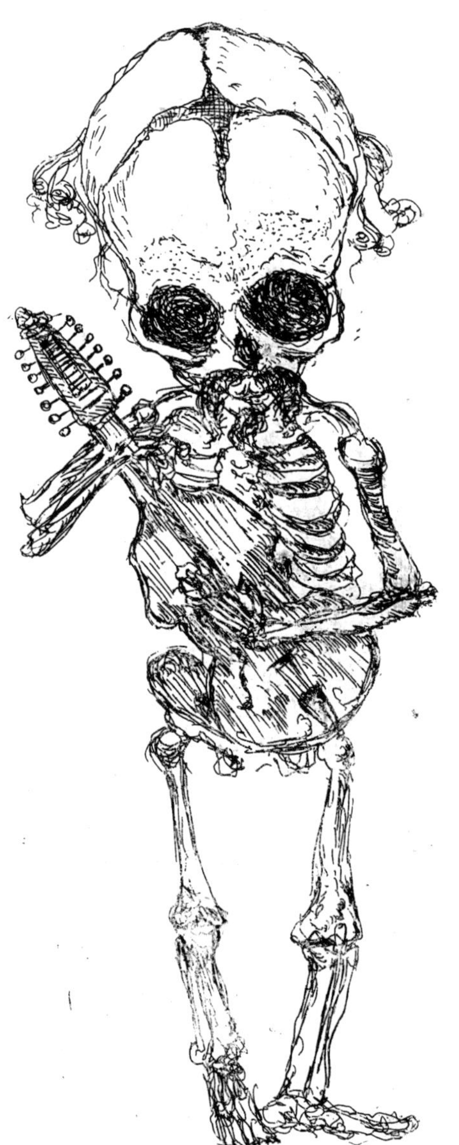

You may say this from us to whoever imagines
it will be a painful effort to breathe his last.

Ruysch. Such reasoning would perhaps satisfy the Epicureans, but not those people who regard the soul as essentially different from the body. I have hitherto been one of the latter, and now that I have heard the dead speak and sing I am more than ever disinclined to change my opinions. We consider death to be a separation of the soul and body, and to us it is incomprehensible how these two substances, so joined and agglutinated as to form one being, can be divided without great force and an inconceivable pang.

Mummy. Tell me: is the spirit joined to the body by some nerve, muscle, or membrane which must be broken to enable it to escape? Or is it a member which has to be severed or violently wrenched away? Do you not see that the soul necessarily leaves the body when the latter becomes uninhabitable, and not because of any internal violence? Tell me also: were you sensible of the moment when the soul entered you, and was joined, or as you say agglutinated, to your body? If not, why should you expect to feel any violent sensation at its departure? Take my word for it, the departure of the soul is as quiet and imperceptible as its entrance.

Ruysch. Then what is death, if it be not pain?

Mummy. It is rather pleasure than anything else. You must know that death, like sleep, is not accomplished in a moment, but gradually. It is true the transition is more or less rapid according to the disease or manner of death. But ultimately death comes like sleep, without either sense of pain or pleasure. Just before death pain is impossible, for it is too acute a thing to be experienced by the enfeebled senses of a dying person. It were more rational to regard it as a pleasure; because most human joys, far from being of a lively nature, are made up of a sort of languor, in which pain has no part. Consequently, man's senses, even when approaching extinction, are capable of pleasure; since languor is often pleasurable, especially when it succeeds a state of suffering. Hence the languor of death ought to be pleasing in proportion to the intensity of pain from which it frees the sufferer. As for myself, if I cannot recall the circumstances of my death, it may be because the doctors forbade me to exert my brain. I remember, however, that the sensation I experienced differed little from the feeling of satisfaction that steals over a man, as the languor of sleep pervades him.

The other Mummies. We felt the same sensation.

Ruysch. It may be as you say, although ever one with whom I have conversed on this subject is of a very different opinion. It is true, however, they have not spoken from experience. Now tell me, did you at the time of death, whilst experiencing this sensation of pleasure, realise that you were dying, and that this feeling was a prelude to death, or what did you think?

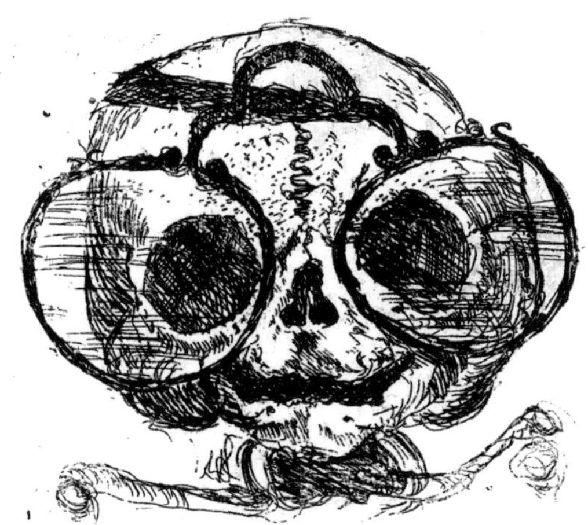

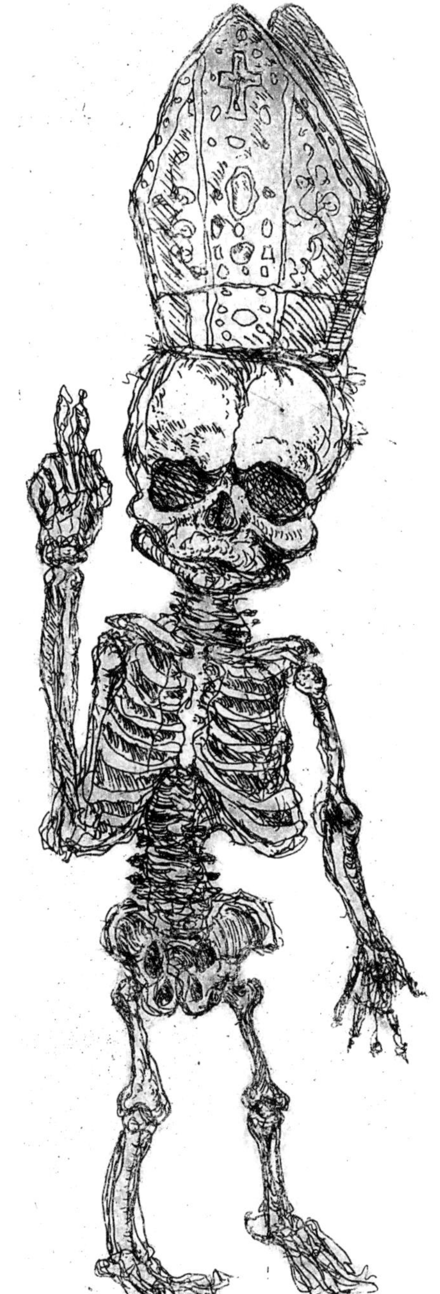

Mummy. Until I was dead I believed I should recover, and as long as I had the faculty of thought I hoped I should still live an hour or two. I imagine most people think the same.

The other Mummies. It was the same with us.

Ruysch. Cicero says that, however old and broken-down a man may be, he always anticipates at least another year of life.

But how did you perceive at length that your soul had left the body? Say, how did you know you were dead? … You do not answer. Children, do you not hear? … Ah, the quarter of an hour has expired. Let me examine them a little. Yes, they are quite dead again. There is no fear that they will give me such another shock. I will go to bed.

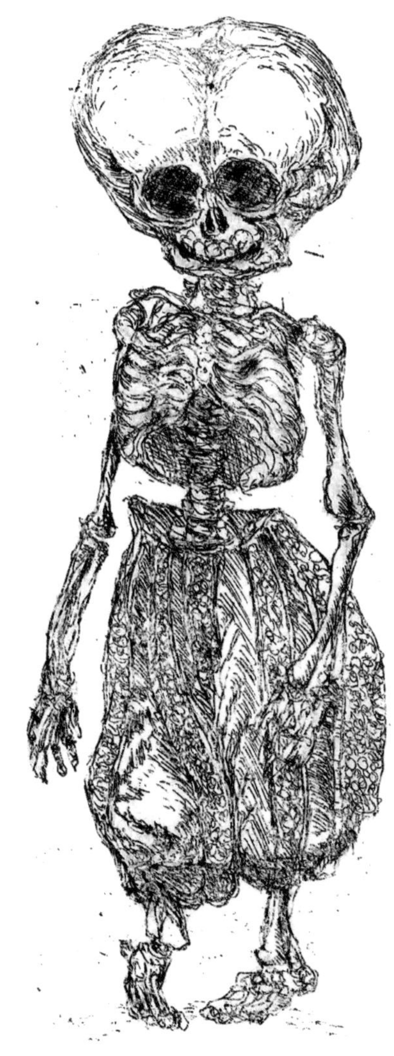

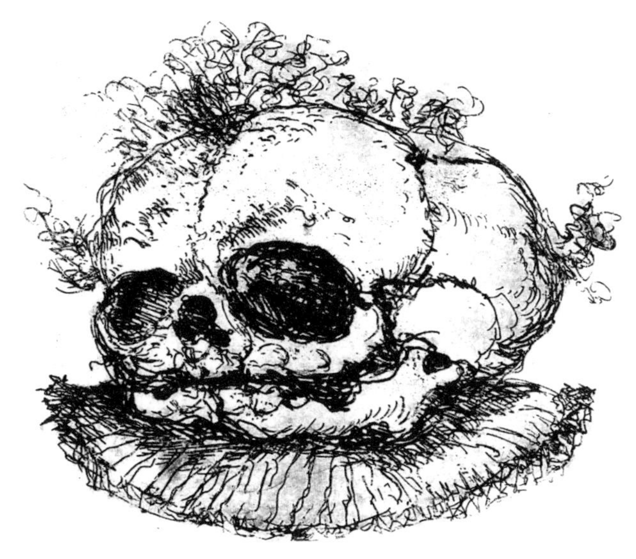

NOTE.

Frederic Ruysch (1638–1731) was one of the cleverest anatomists Holland has ever produced. For sixty years he held a professorship of anatomy at Amsterdam, during which time he devoted himself to his art. He obtained from Swammerdam his secret of preserving corpses by means of an injection of coloured wax. Ruysch, it is said, also made use of his own blood for this purpose. His subjects, when prepared, looked like living beings, and showed no signs of corruption. Czar Peter visited Holland in 1698, and was amazed at what he saw in Ruysch's studio. In 1717 the Czar again visited Holland, and succeeded in inducing Ruysch to dispose of his collection of animals, mummies, &c. These were all transported to St. Petersburg. Ruysch formed a second collection as valuable as the first, which after his death was publicly sold.

Specimens and Other Ruyschiana

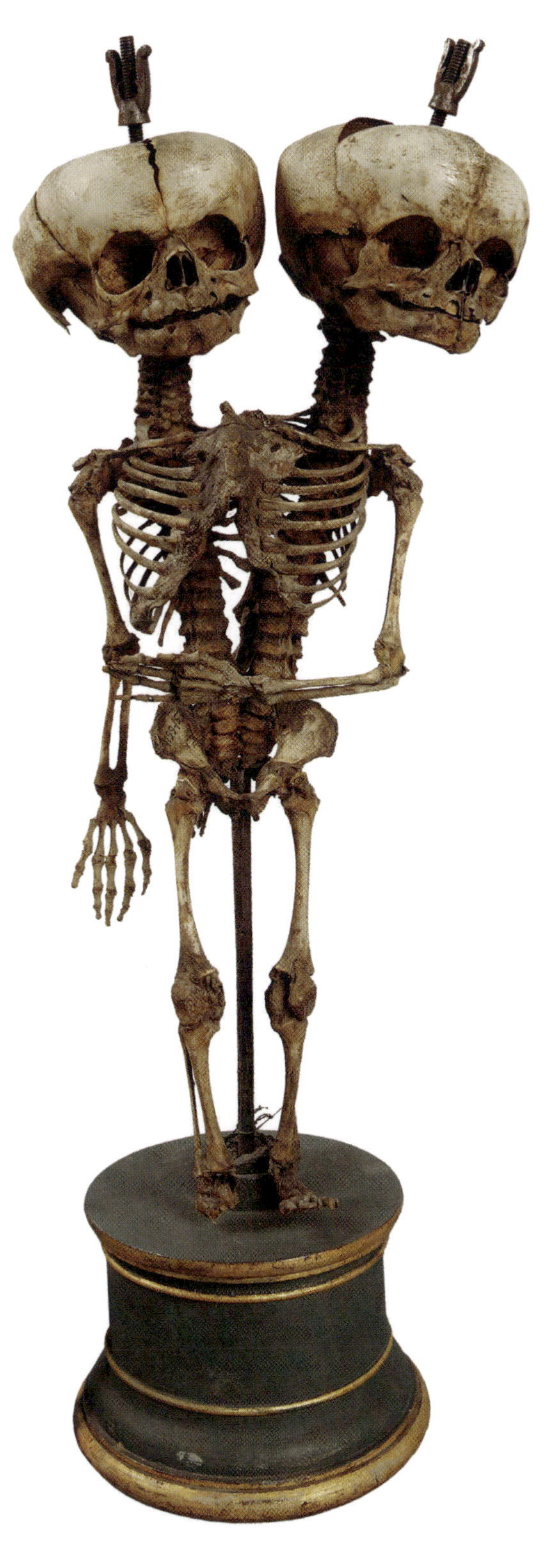

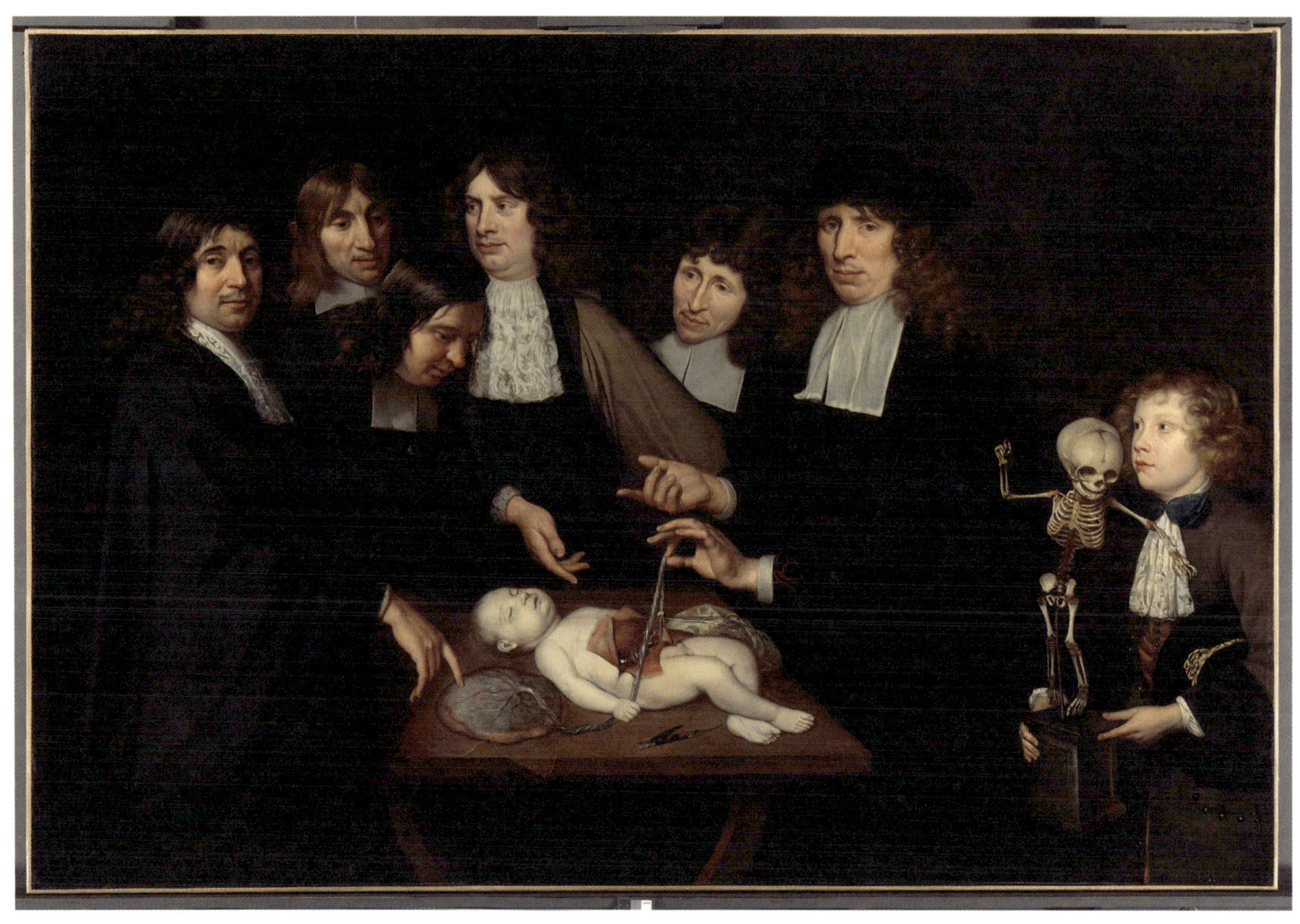

Above: The Anatomy Lesson of Dr. Frederik Ruysch. *Jan van Neck, 1683. This painting is part of a larger group of portraits of the Amsterdam Guild of Surgeons at Amsterdam Museum. Courtesy of Amsterdam Museum*

Facing page: Child's head in a turban resting on a pillow, an unattributed specimen dated around 1720 and thought to have been created by Frederik Ruysch, in the collection of the Museum Bleulandinum, the Anatomical Museum of the Utrecht University Medical Center, The Netherlands. Photographer: Peter Rothengatter, HetFotoAtelier, Utrecht

Previous page: Ruysch preparation of a child's skeleton with two heads and three arms, late seventeenth–early eighteenth century. Part of the collection acquired by Peter the Great for his Kunstkammer. *Peter the Great Museum of Anthropology and Ethnography (Kunstkamera) of the Russian Academy of Sciences*

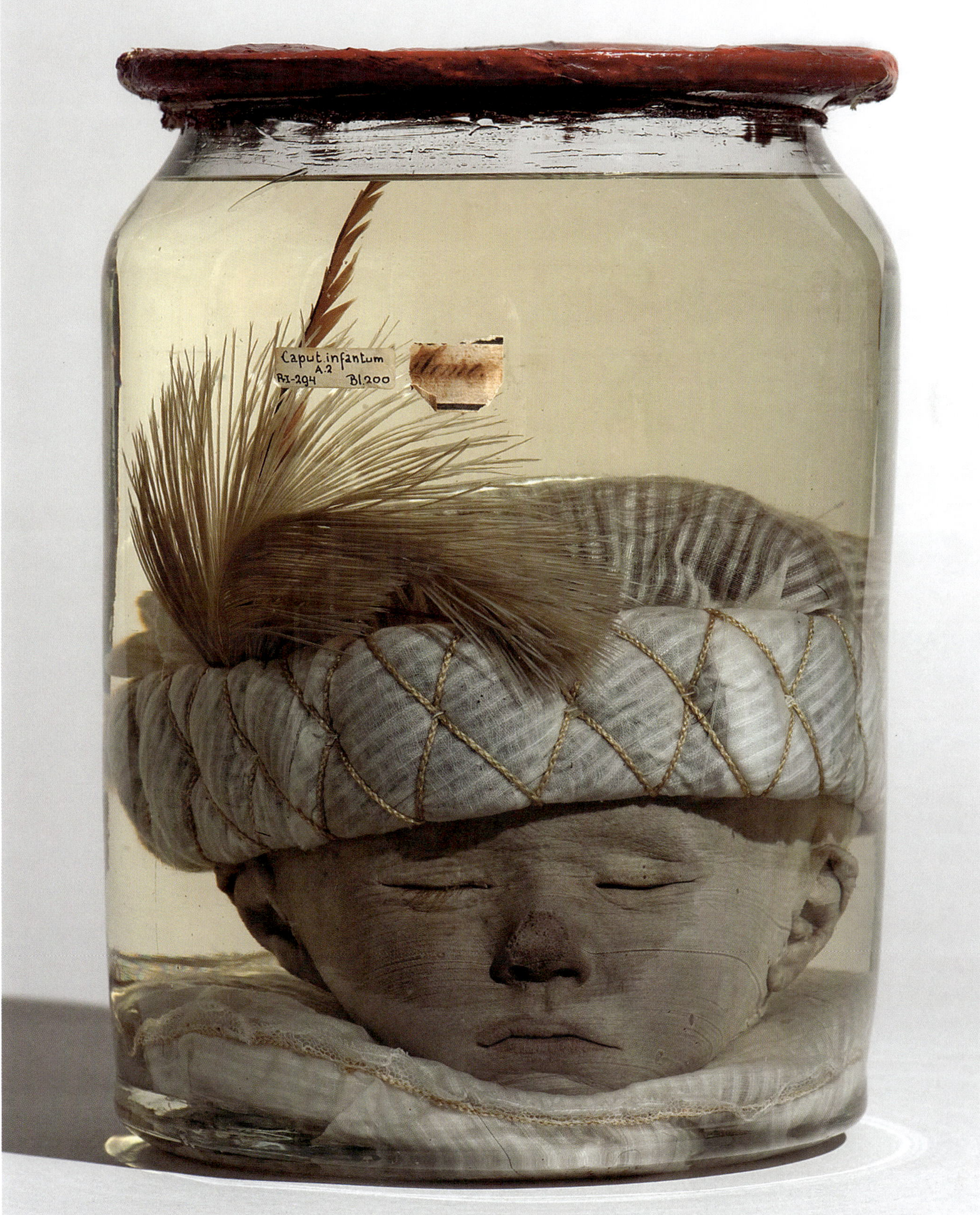
Caput infantum
A.2
RI-294 Bl.200

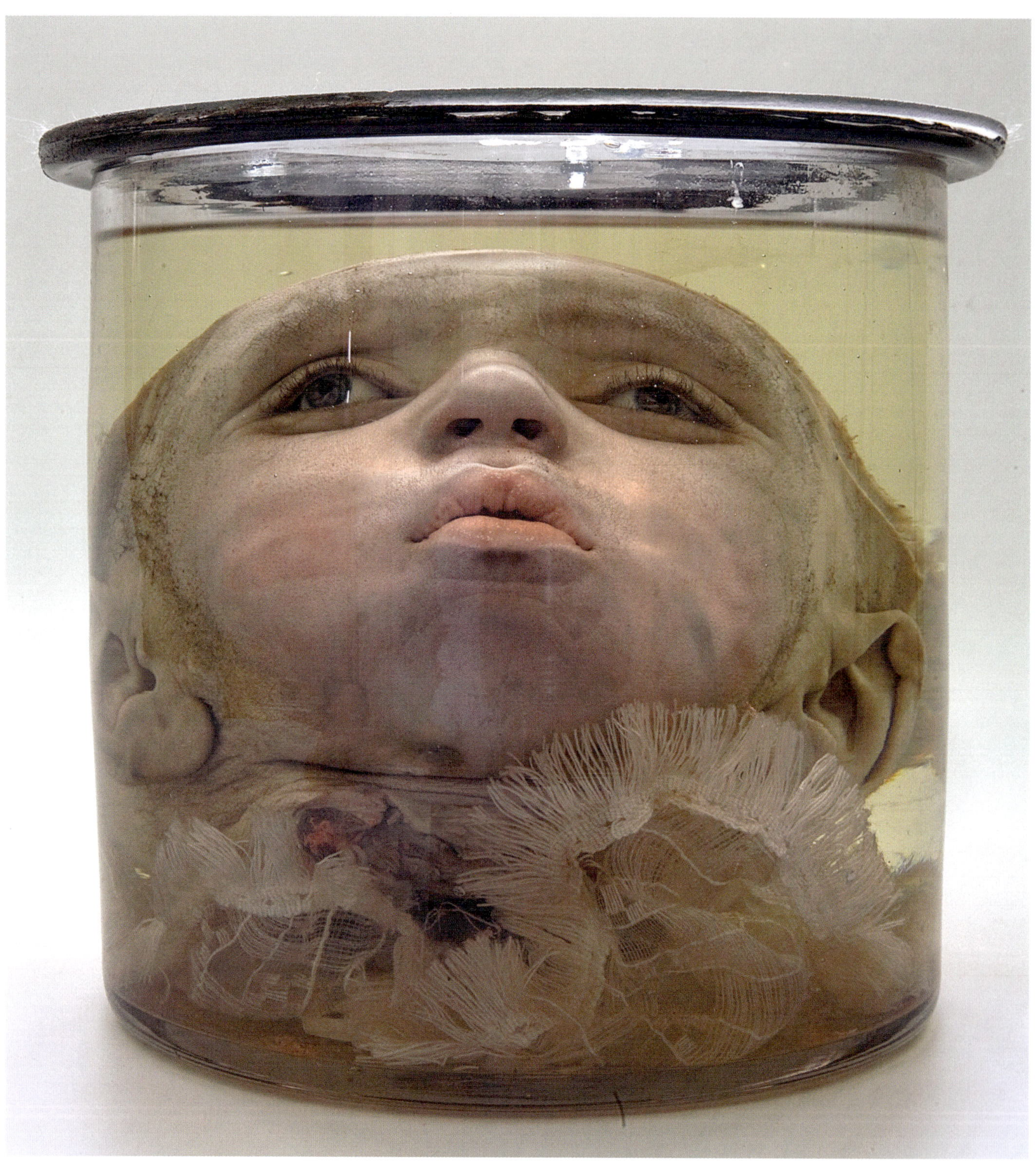

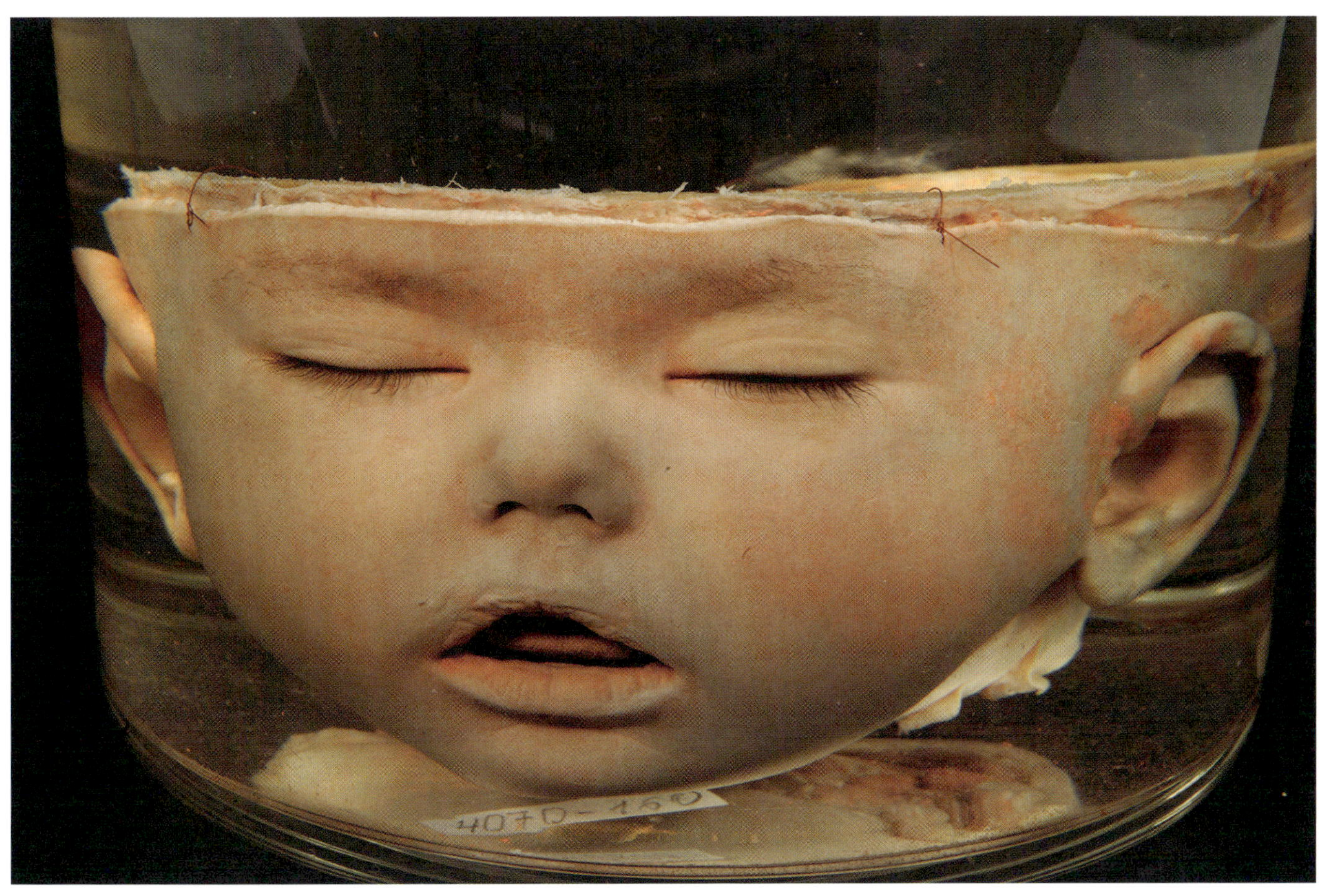

Facing page: Head of young child, prepared by Frederik Ruysch, late seventeenth–early eighteenth century. Courtesy of the Peter the Great Museum of Anthropology and Ethnography (Kunstkamera) of the Russian Academy of Sciences

Above: Injected head of a child with dissected cranium, prepared by Frederik Ruysch, late seventeenth–early eighteenth century. Courtesy of the Peter the Great Museum of Anthropology and Ethnography (Kunstkamera) of the Russian Academy of Sciences

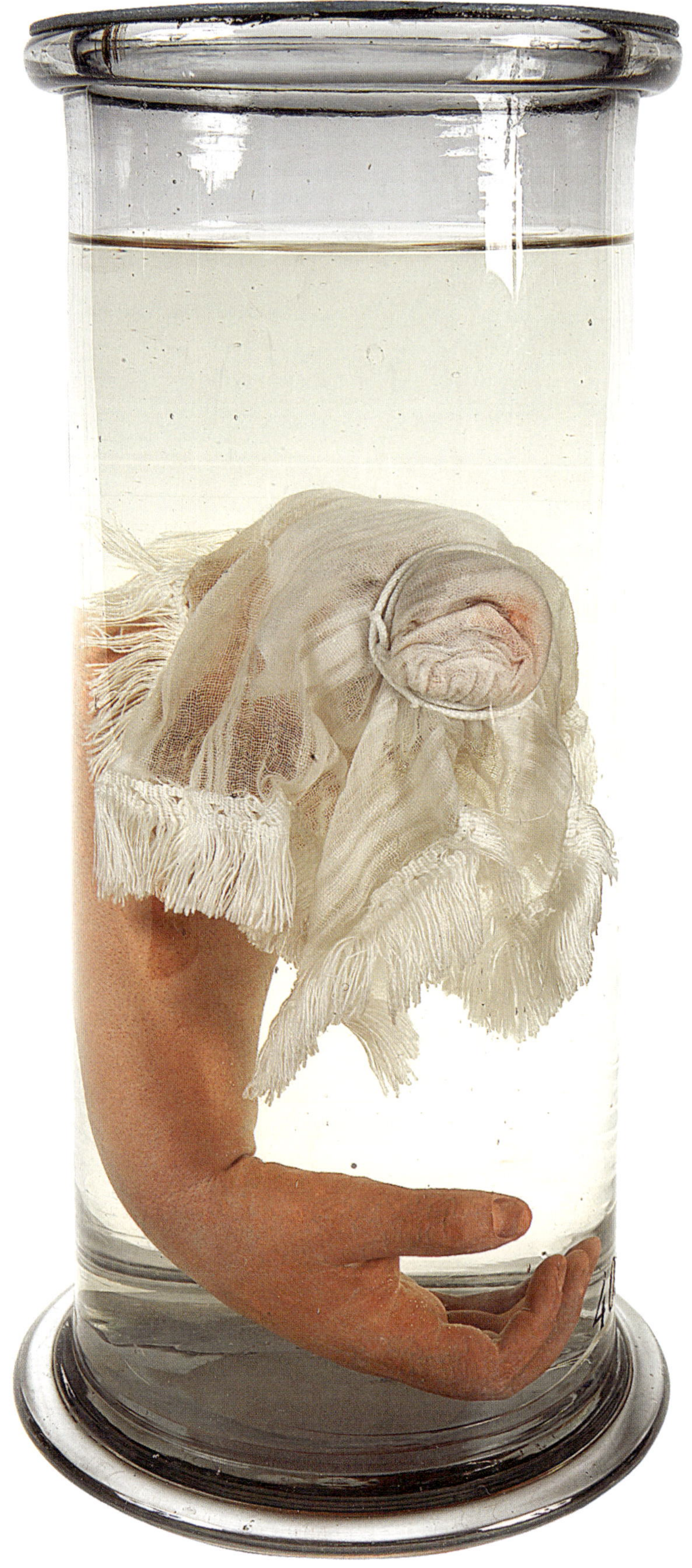

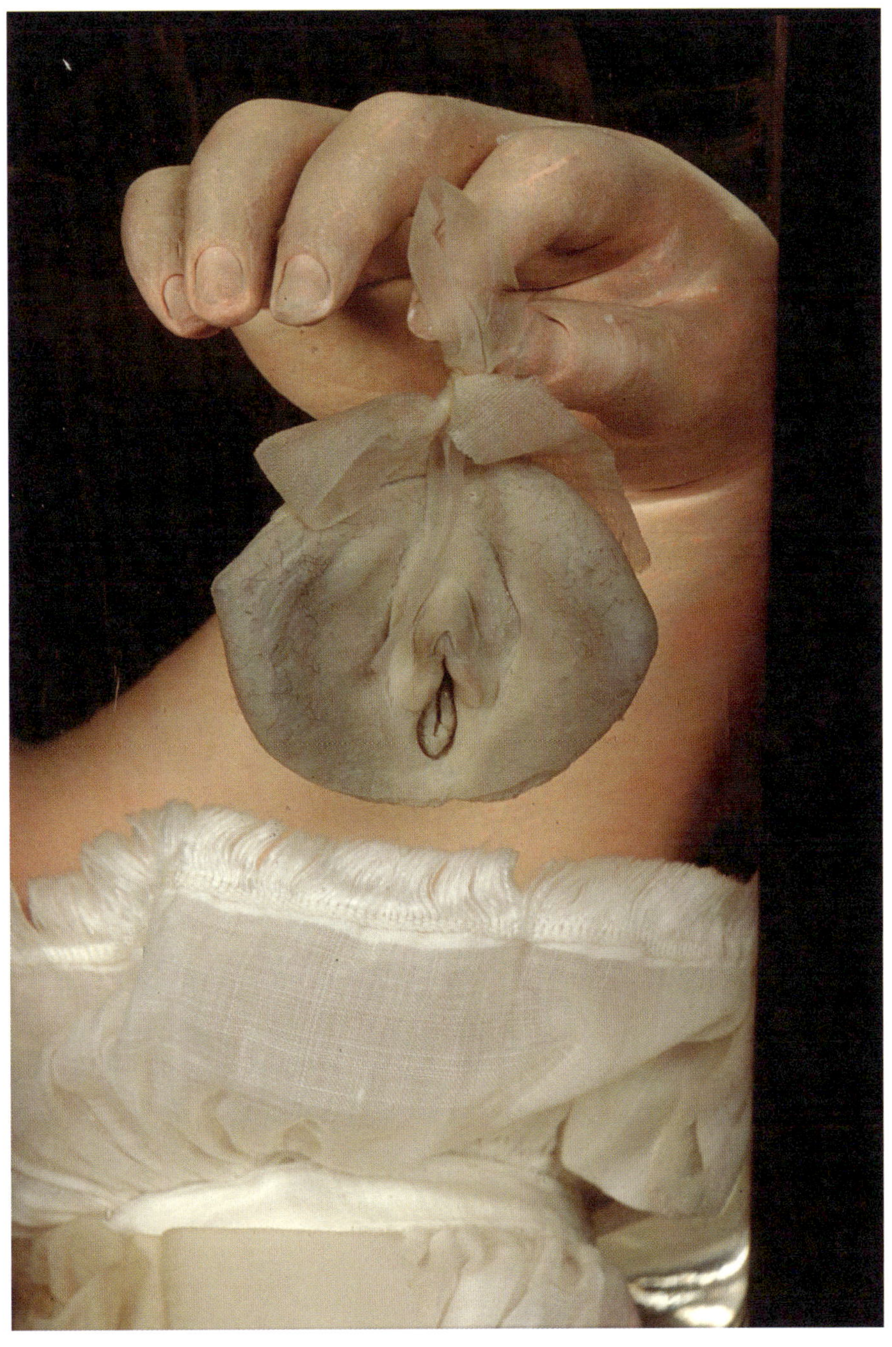

Facing page: Preparation of child's arm, prepared by Frederik Ruysch, late seventeenth–early eighteenth century. Courtesy of the Peter the Great Museum of Anthropology and Ethnography (Kunstkamera) of the Russian Academy of Sciences

Above: Frederik Ruysch preparation consisting of an infant's arm holding a vulva dangling from a ribbon, from the collection of the Leiden University Medical Centre (LUMC). Courtesy of Leiden University Medical Centre (LUMC)

Following pages: Fetus with injected umbilical cord (left) and preparation of child's leg with scorpion(right), prepared by Frederik Ruysch, late seventeenth–early eighteenth century. Courtesy of the Peter the Great Museum of Anthropology and Ethnography (Kunstkamera) of the Russian Academy of Sciences

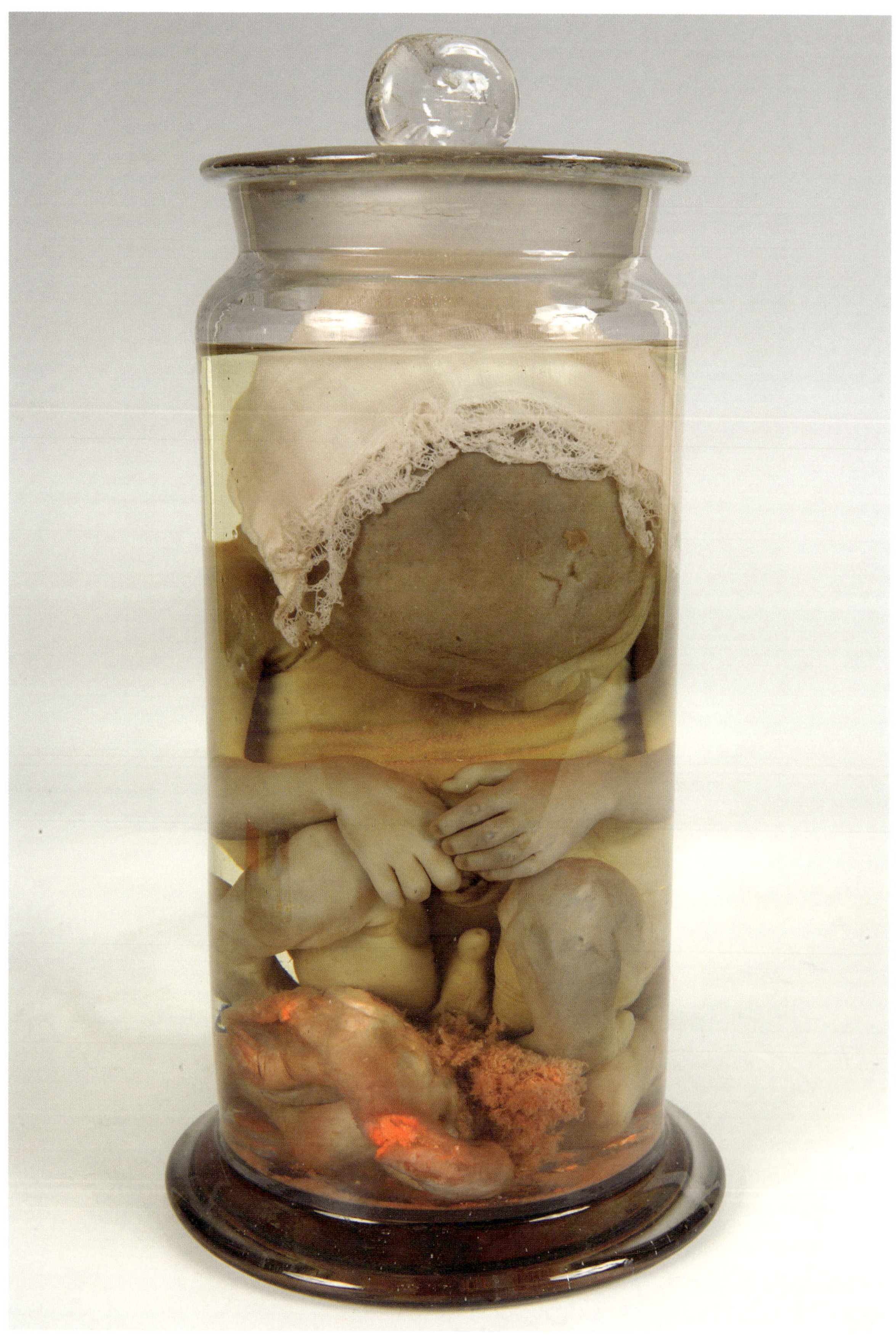

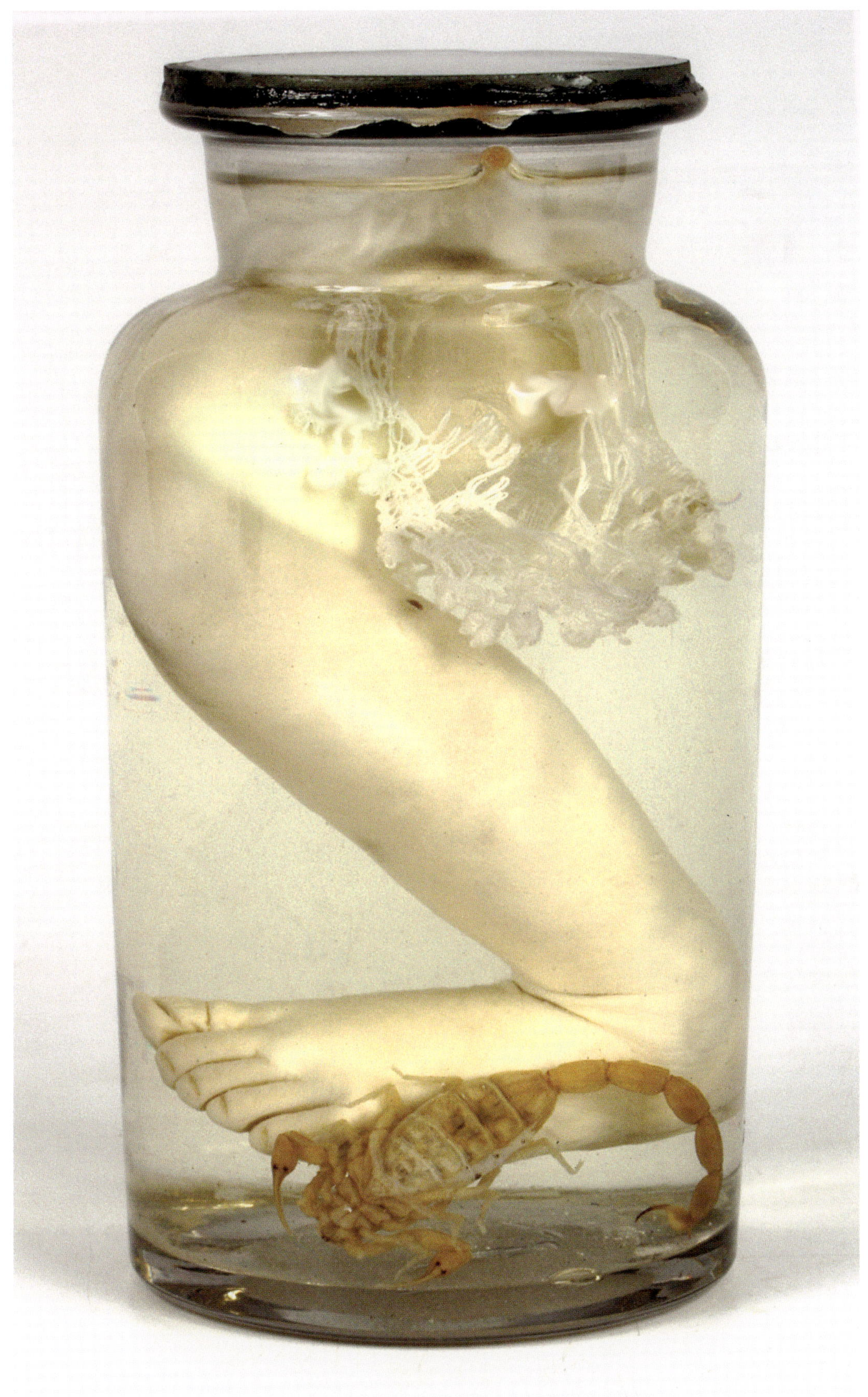

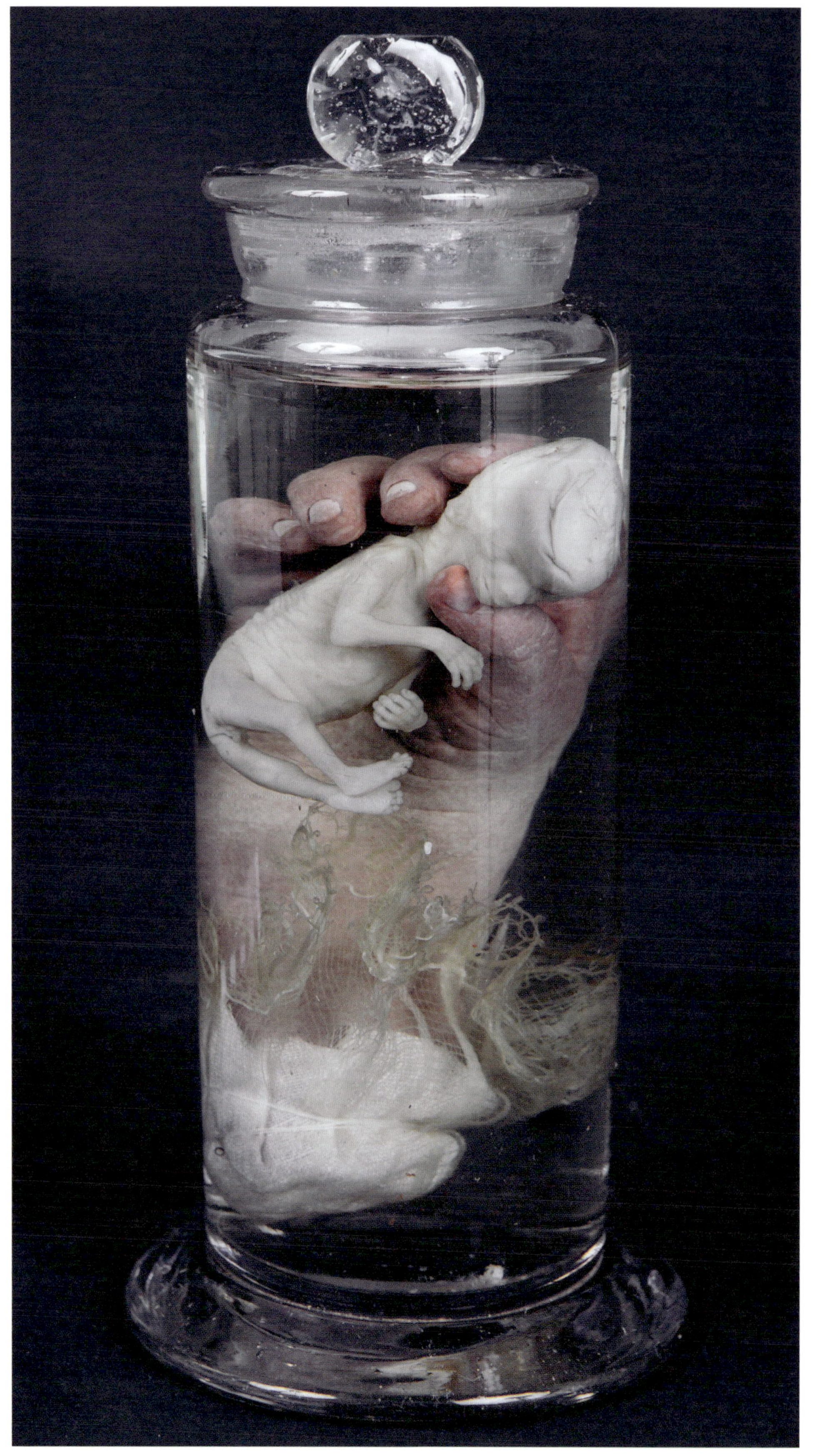

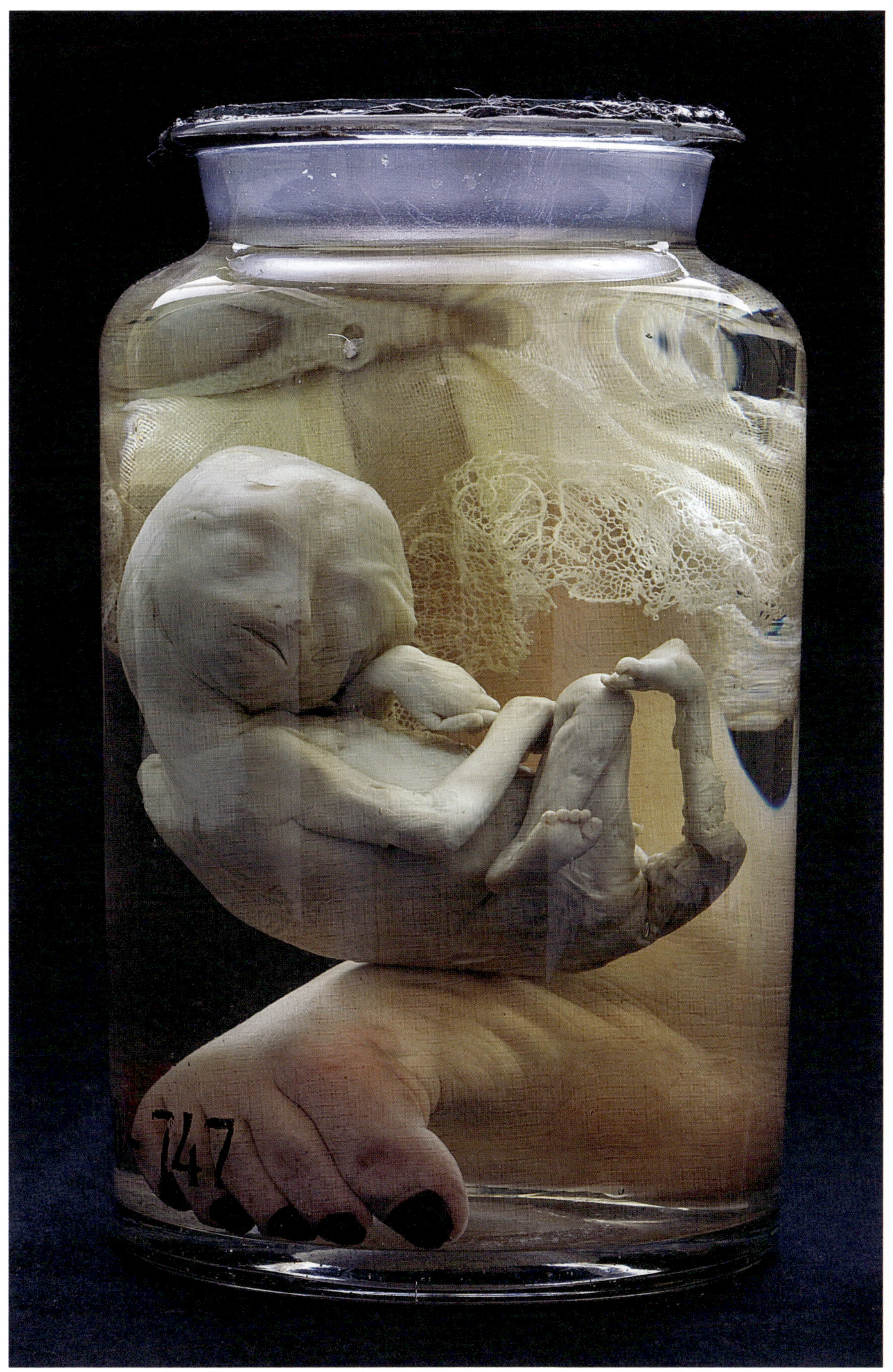
747

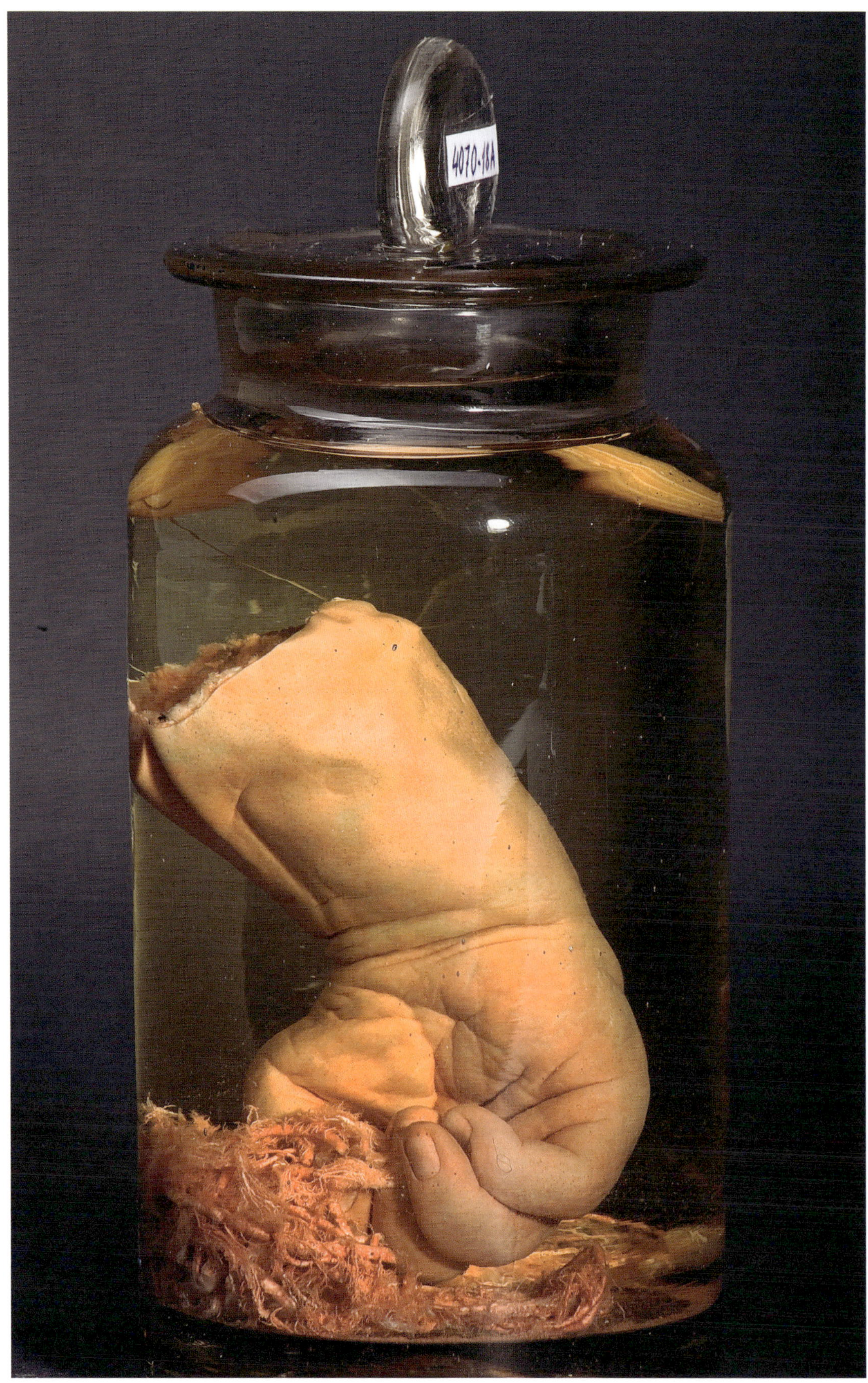
4070-18A

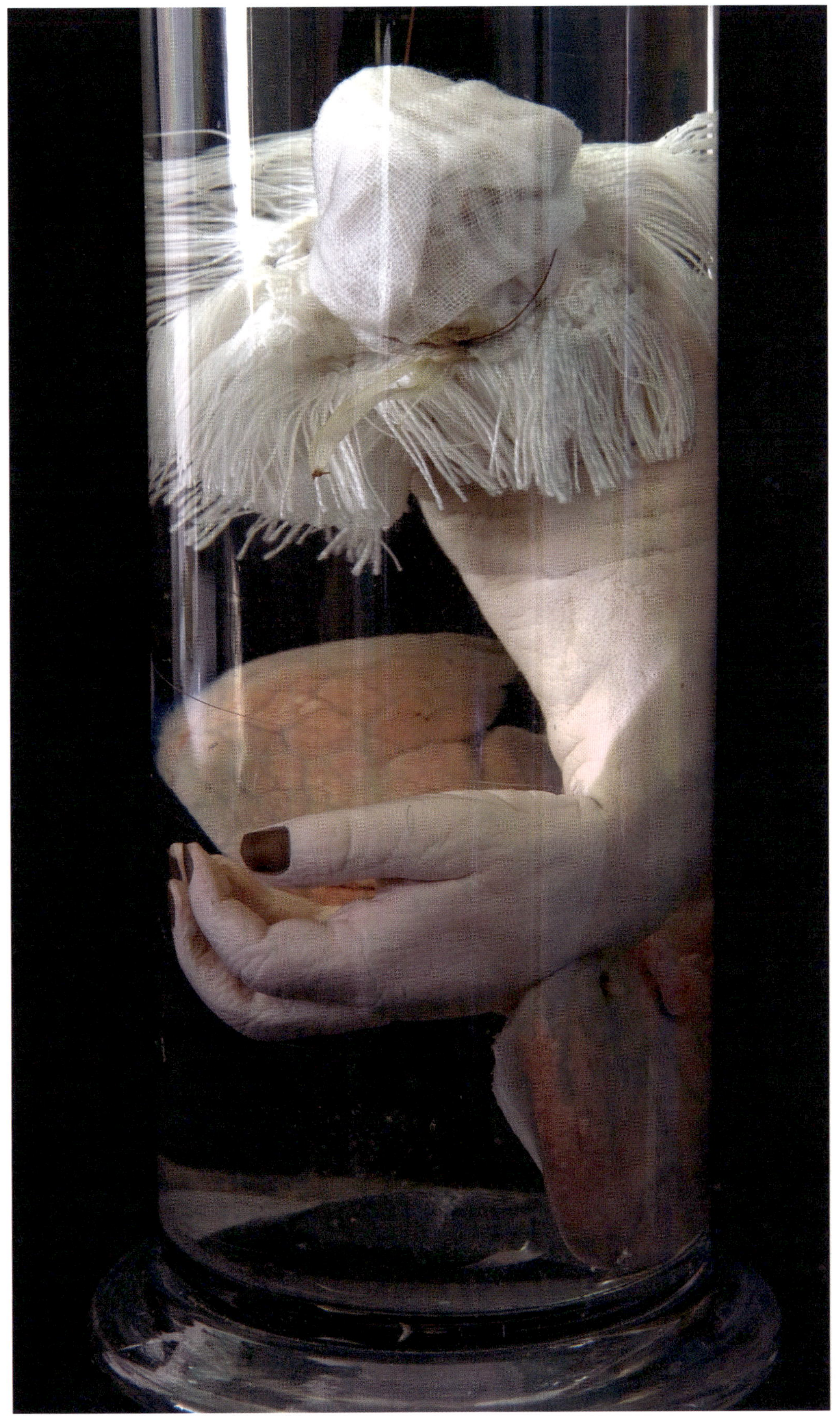

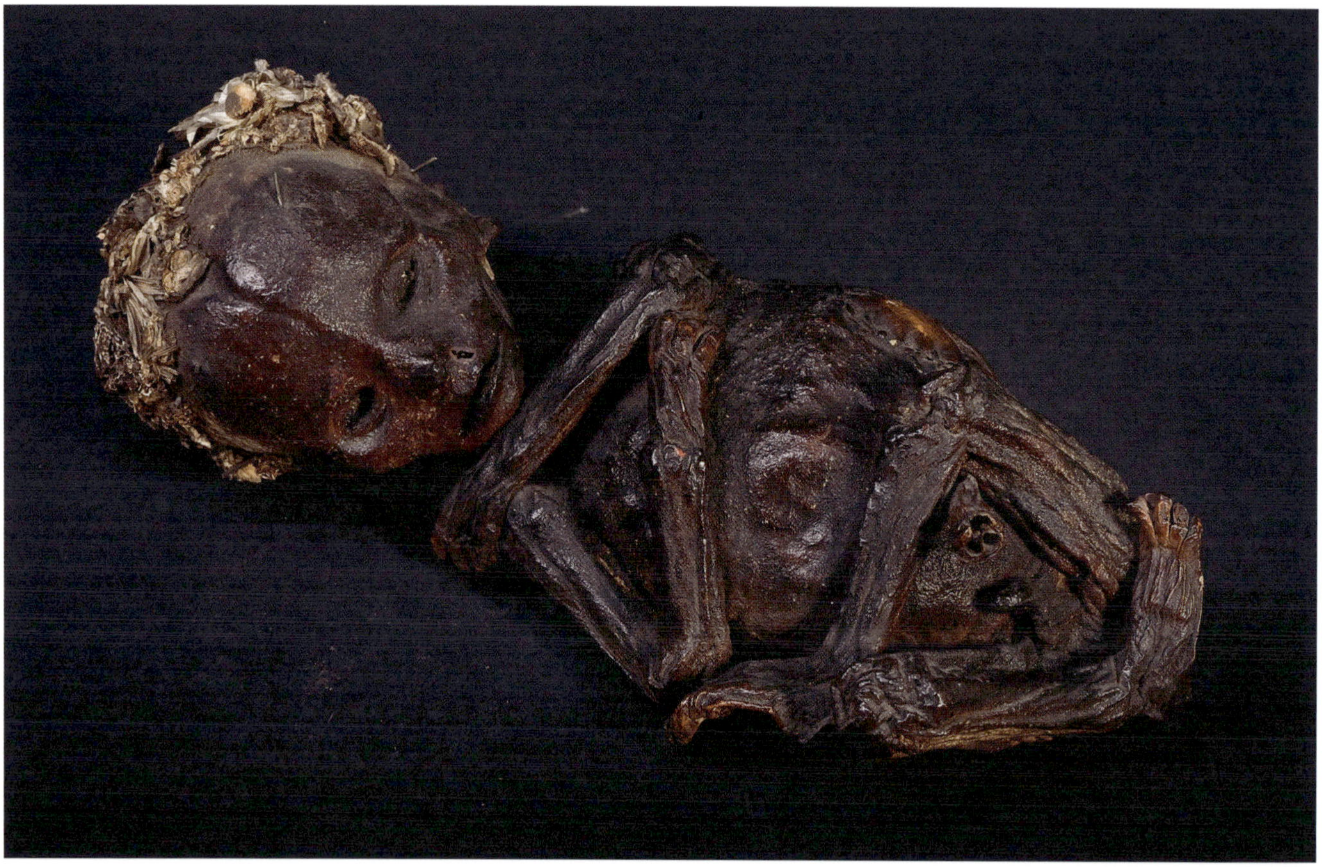

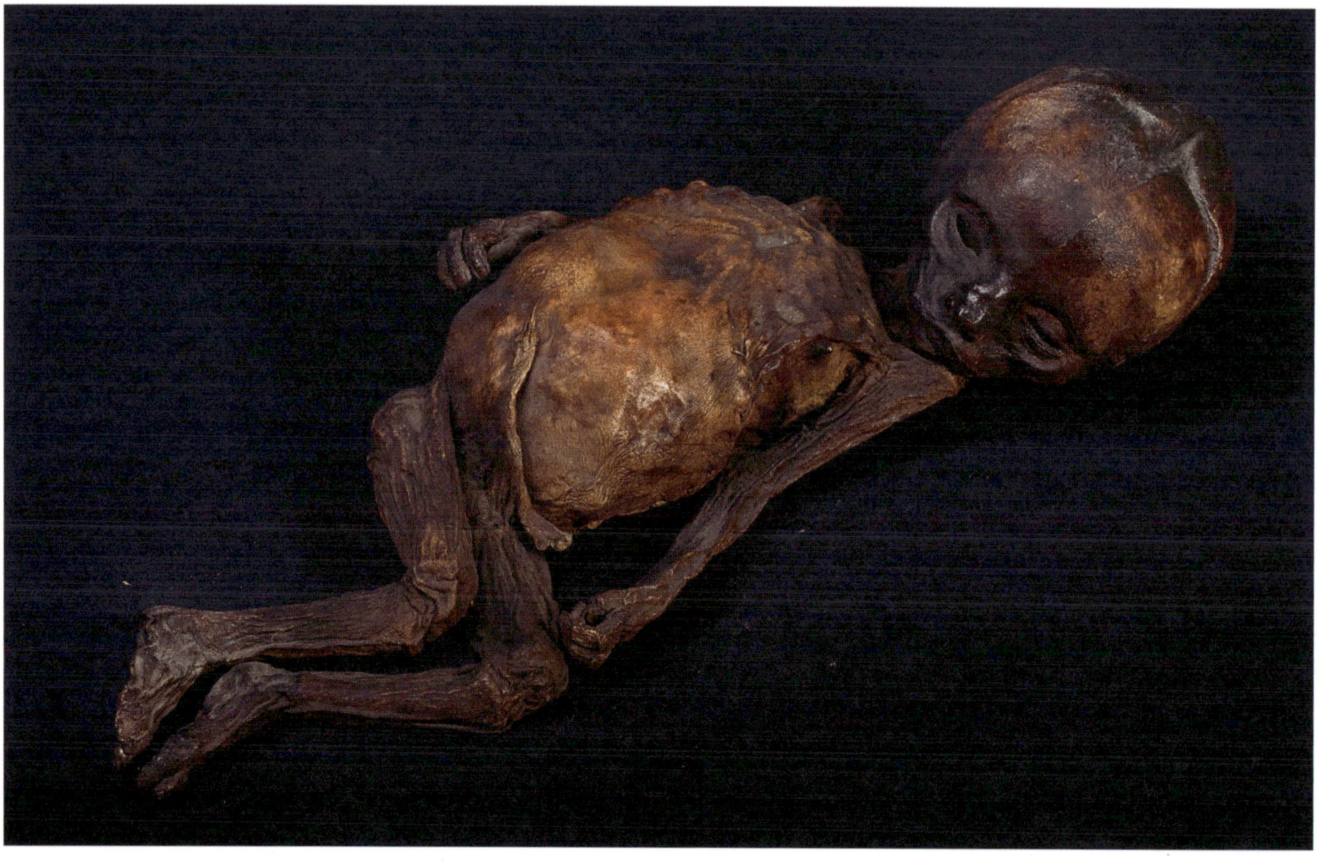

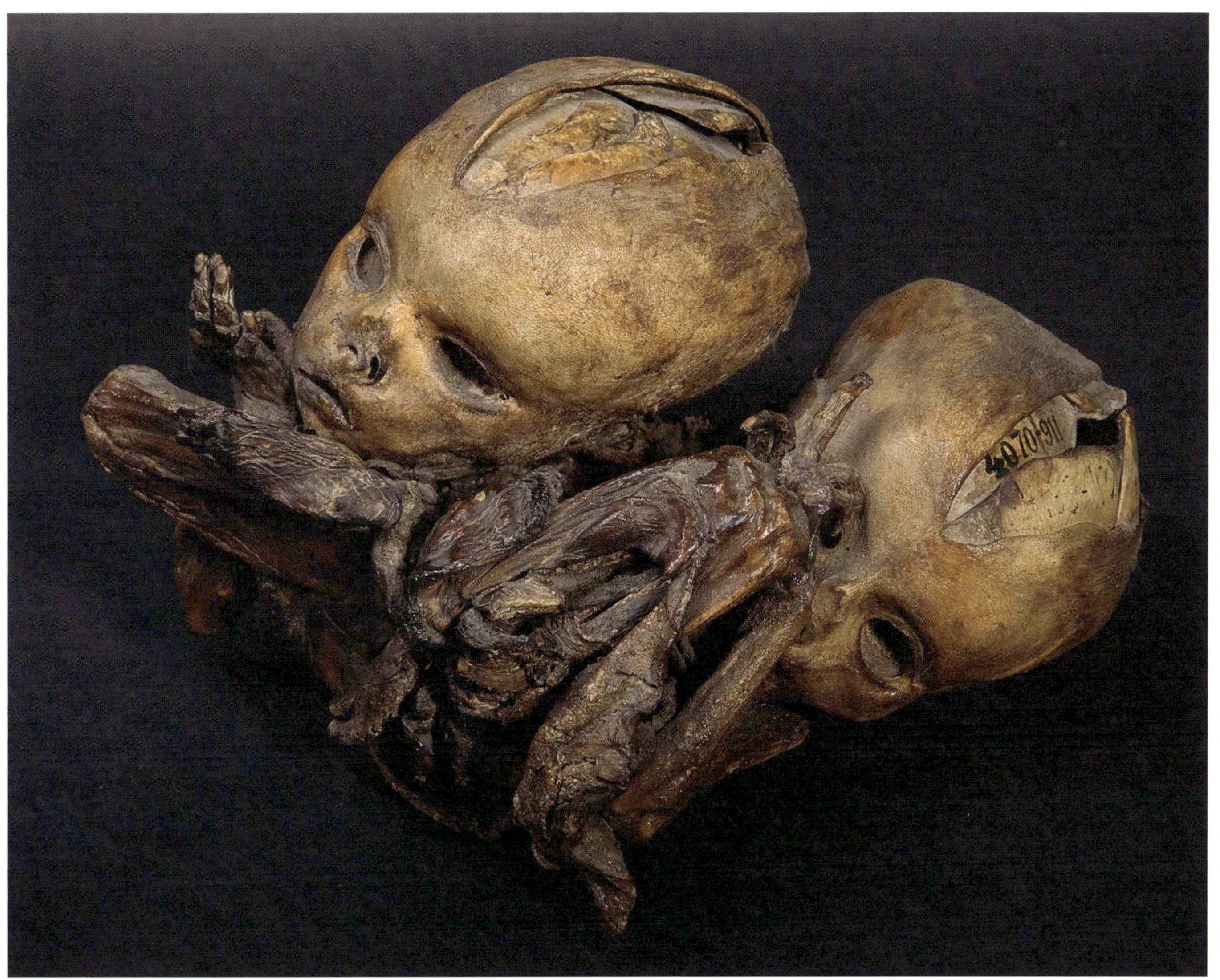

Pages 94–95: Fetus about two months old in the hand of child (left) and three-month-old fetus with injected leg (right). Pages 96–97: Right hand of a one-year-old child (left) and injected lung of baby fixed to the hand of child (right). Prepared by Frederik Ruysch, late seventeenth–early eighteenth century. Courtesy of the Peter the Great Museum of Anthropology and Ethnography (Kunstkamera) of the Russian Academy of Sciences

Left: Embalmed fetus about six months old with flower wreath on head (top) and embalmed fetus about eight months old (bottom) prepared by Frederik Ruysch. Courtesy of the Peter the Great Museum of Anthropology and Ethnography (Kunstkamera) of the Russian Academy of Sciences

Above: Two-headed newborn in sitting position prepared by Frederik Ruysch. Courtesy of the Peter the Great Museum of Anthropology and Ethnography (Kunstkamera) of the Russian Academy of Sciences

Next four pages: Illustrations from other publications by Frederik Ruysch. Page 100: illustration showing wax injection of the blood vessels in the anachoid and pia mater, from Opera omnia anatomica- medico- chirurgica ... *(1721); remaining images illustrating Ruysch specimens from* Thesaurus animalium primus *(Amsterdam: Joannem Wolters, 1710)*

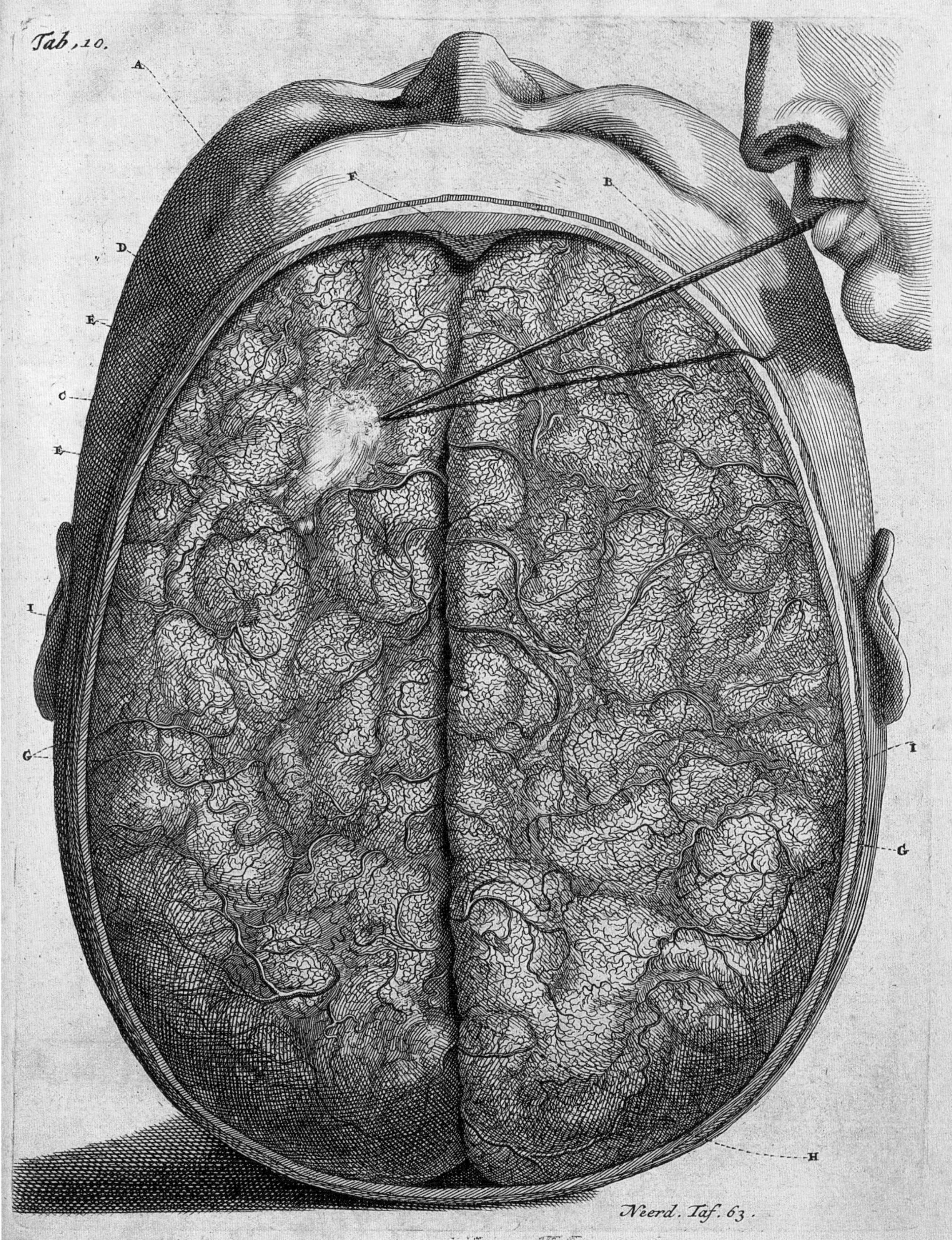
Tab, 10.
A
B
F
D
E
C
E
I
G
I
G
H
Neerd. Taf. 63.

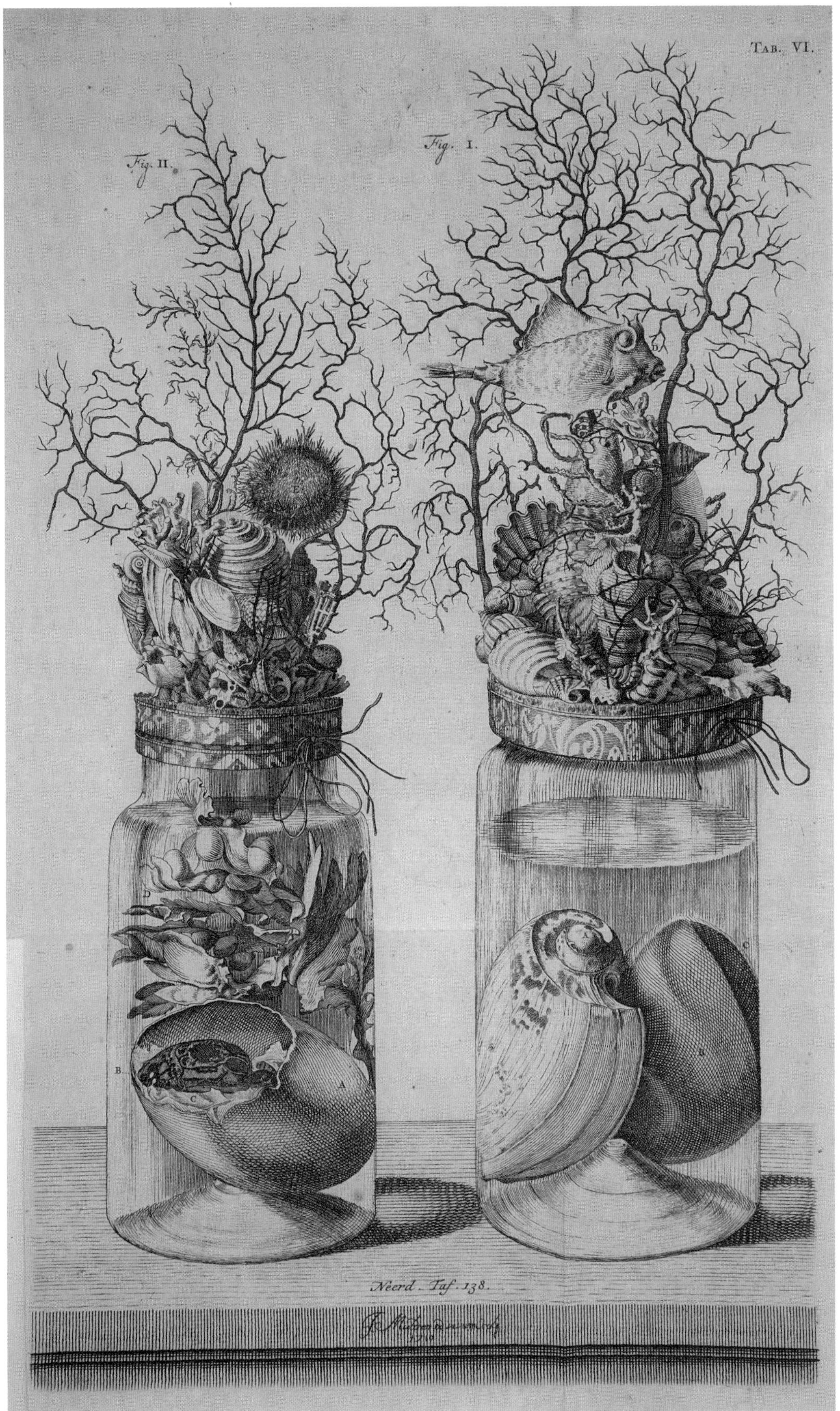
Tab. VI.
Fig. II.
Fig. I.
Neerd. Taf. 138.

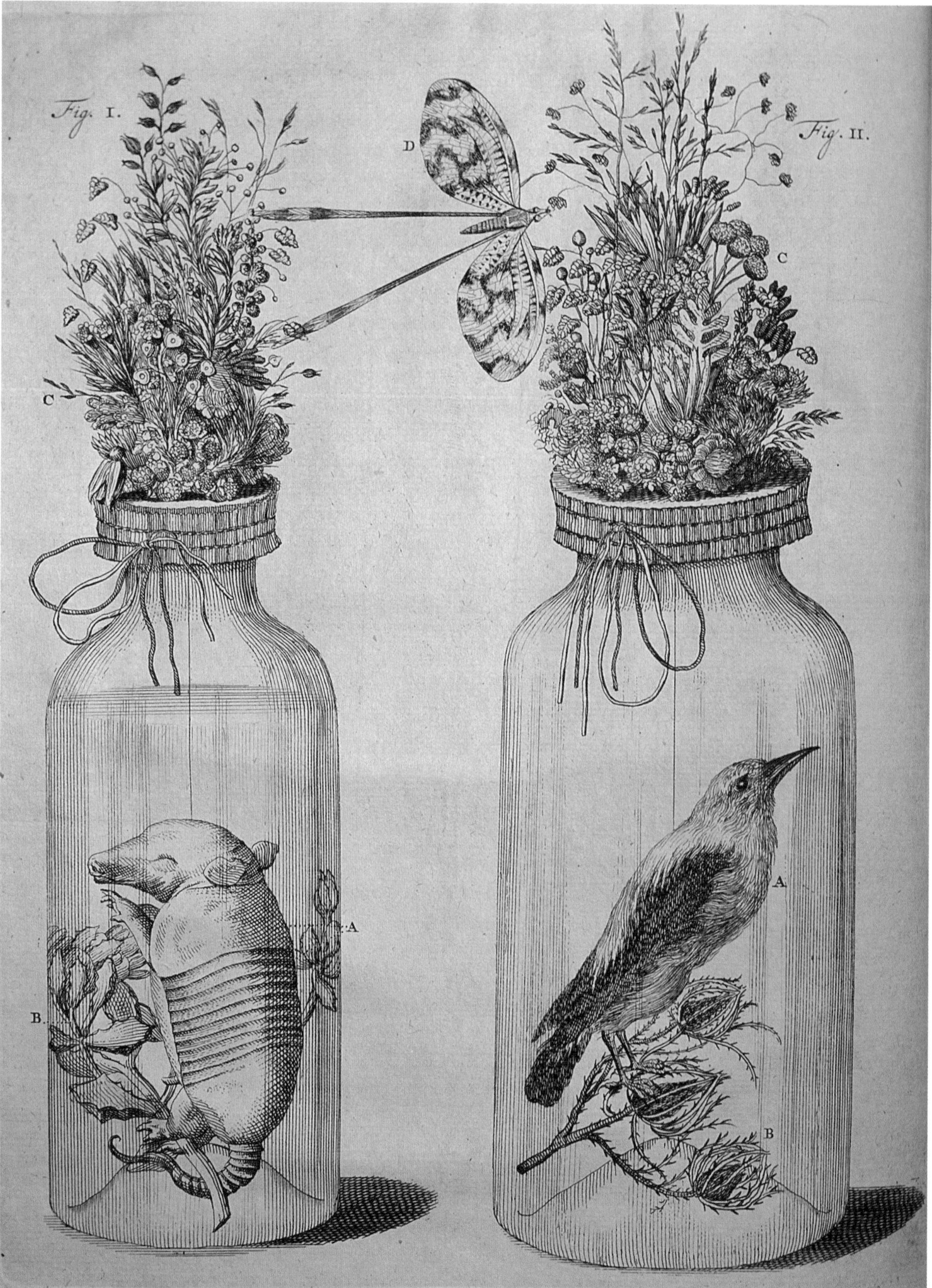

Fig. I.
D
Fig. II.
C
C
A
B.
A.
B

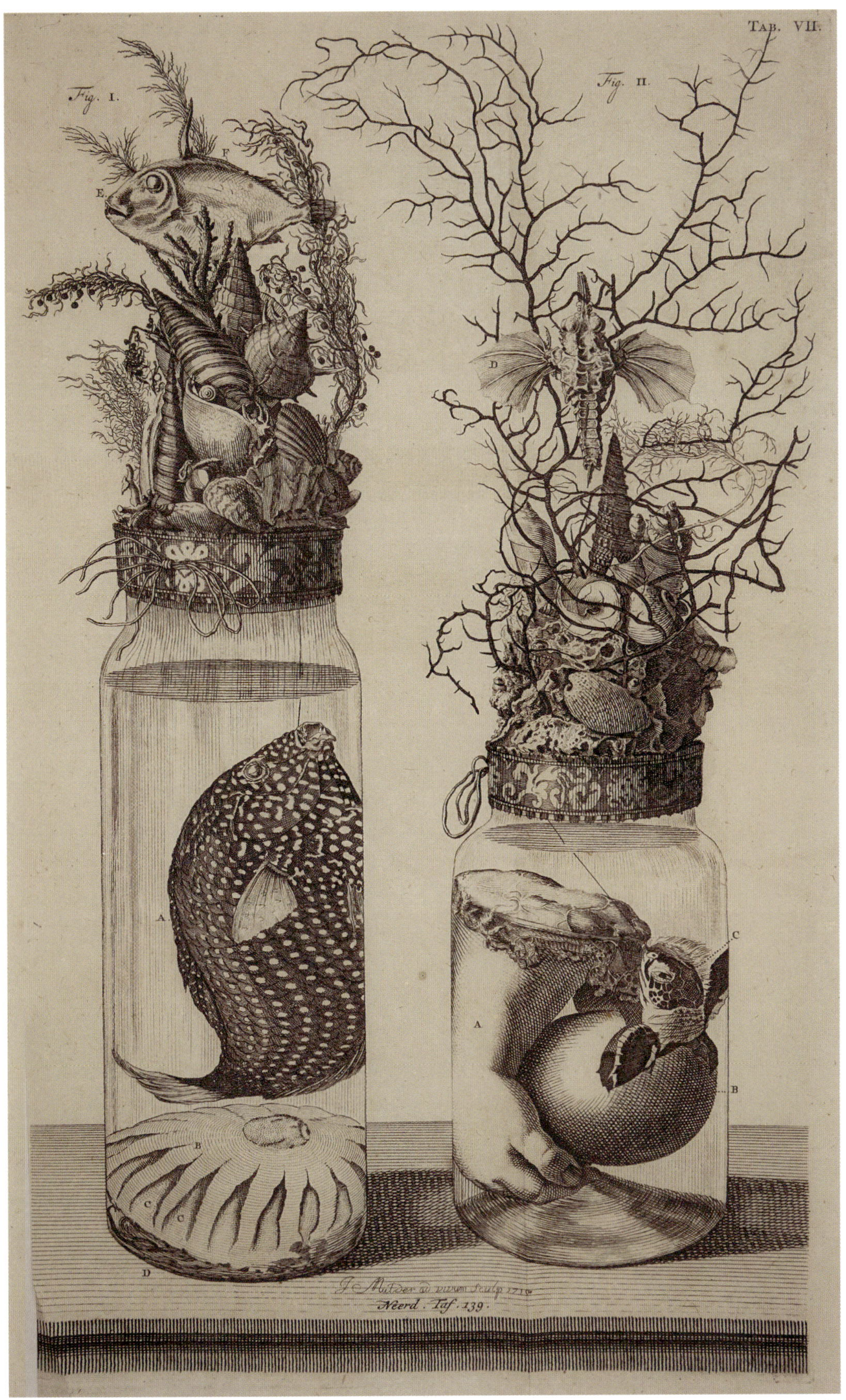

TAB. VII.
Fig. I.
Fig. II.
E.
F
D
A
B
C
C
C
D
A
B
C
J. Mulder ad vivum sculp. 1710
Neerd. Taf. 139.

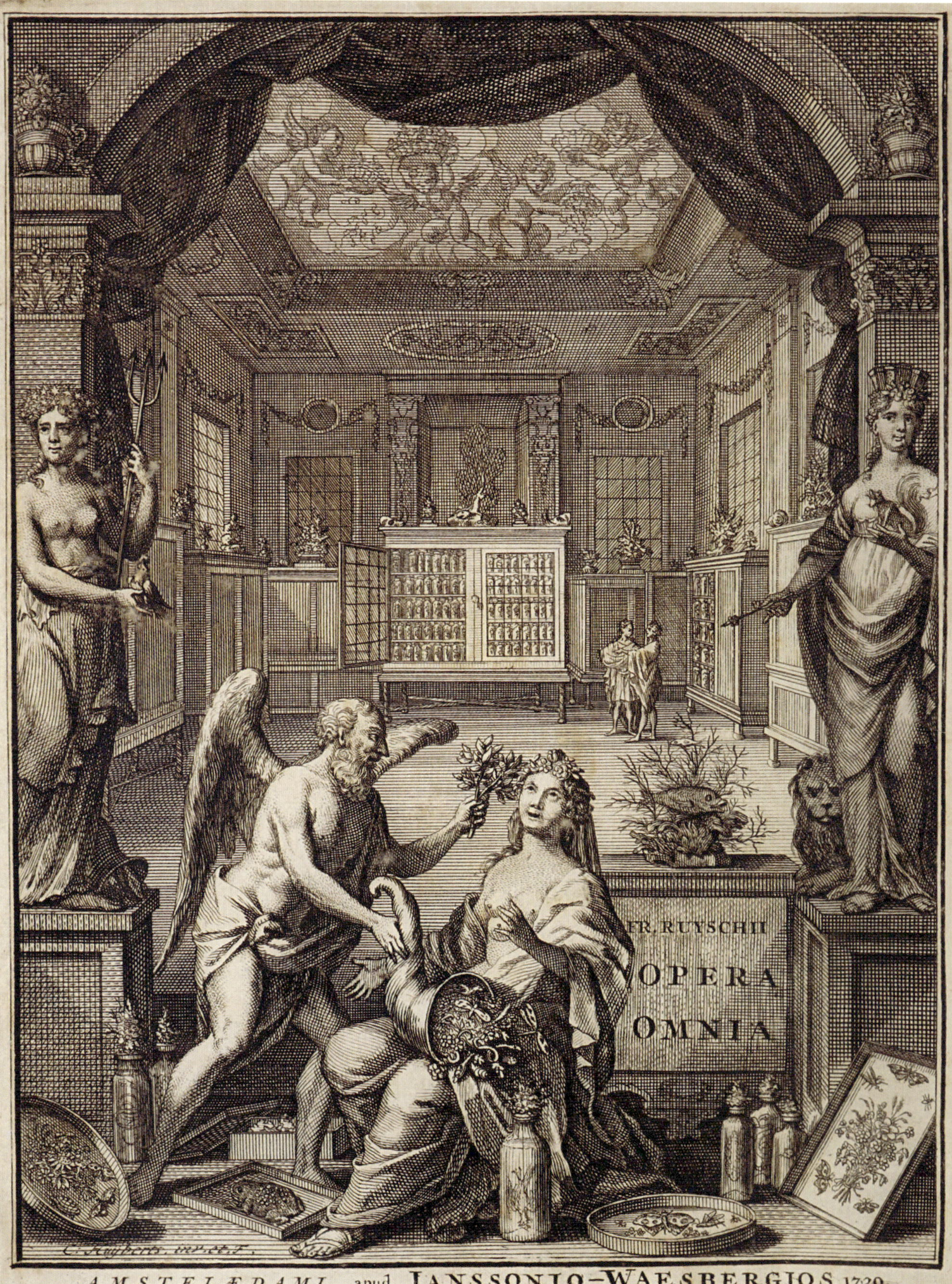
FR. RUYSCHII
OPERA
OMNIA
AMSTELÆDAMI apud JANSSONIO-WAESBERGIOS 1720.

ABRIDGED TRANSLATION OF FREDERIK RUYSCH'S

THESAURUS ANATOMICUS

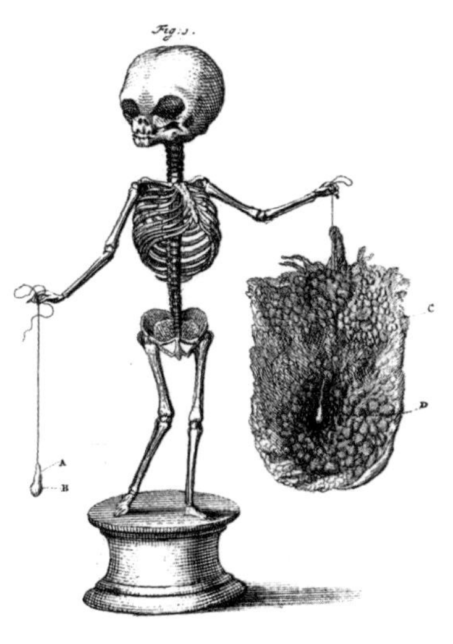

translated by Richard Faulk

Translator's Introduction

BY RICHARD FAULK

BY ANY STANDARD, FREDERIK RUYSCH'S THESAURUS ANATOMICUS IS a deeply weird work. Published in ten volumes, in Latin and Dutch, between 1701 and 1716, the series has been received by the modern English-speaking world as essentially a picture book of medical oddities, interrupted by stretches of gorgeous and impenetrable eighteenth-century text. Deciphering that text only deepens the overall strangeness. In pictures and words, the hallmark of the *Thesaurus anatomicus* is a surreal juxtaposition of content: dry-as-dust medical observations punctuated by hexameter verses and moral reflections; jars of painstakingly preserved anatomical specimens interspersed with fanciful landscapes of gallstone hills ascended by baby skeletons; a severed arm swaddled in linen and lace; a fetus kept in a jar enlivened with a sash of blue silk.

This vision is uniquely Ruyschian, but it is not pathological. The attention that this gifted anatomist and preparator lavished on the aesthetics of his medical specimens was not unremarked on by his contemporaries, but the consensus was mostly positive.[45] Certainly they saw nothing ghoulish about it. Famously, Peter the Great (1672–1725) kissed the lips of one of Ruysch's preserved infants, moved to tears by its pathetic beauty. Broadly speaking, this was the response Ruysch was hoping to elicit. For in opening to the public his private anatomical collection (which was housed in a series of massive display cabinets, kept in three separate rooms of his house), Ruysch intended to raise the profile of medical research—and, of course, of his role within it. The subsequent publication of the *Thesaurus anatomicus*, the illustrated catalogue of his museum, was a bid to reach an even larger audience. Ruysch had no illusions about the gloomy reputation that dogged his profession, and he frankly admitted that it was a dirty, smelly, and tedious business. So, while he doubtless had personal reasons for taking so much delight in ensuring the lifelike appearance of his specimens,[46] his efforts were also part of a clear-sighted public relations strategy. Indeed, the final volume of the *Thesaurus anatomicus* begins with a sales pitch. Ruysch, believing he was near the end of his life, wanted to unload his entire collection, and he wanted top dollar. There was no dividend in frightening away potential customers.

Image on page 104: Frederik Ruysch's cabinet, as depicted in the frontispiece for Opera omnia anatomica-medico-chirurgica ... *(1721).*

However well we may understand Ruysch's historical context or his intentions, his collection nevertheless retains an irreducible strangeness that doesn't fail to startle. It's this weirdness that preserved a cult following for the *Thesaurus anatomicus* long after its scientific usefulness expired. Generations have turned to its illustrations—charming, eerie, inscrutable—to gawk, to dream, to ponder. It is an interest that Ruysch himself never dismissed as prurient. Rather, he welcomed the gawkers, confident that they would leave at least marginally improved, intellectually and morally, by the contemplation of his life's work.

It is with this audience in mind that I have prepared this translation. I hope that academics and those whom Ruysch might have called "enthusiasts of natural philosophy" will find something of value here. But above all, this was written for nonexperts who might find in Ruysch's strange and imaginative project inspiration for their own creativity.

The effects of this intention are felt less in *how* I have translated than in *what* I have chosen to translate. Each of the ten volumes of the original *Thesaurus* consists of two parts: the first is a detailed enumeration of the contents of one of Ruysch's display cabinets; the second consists of full-page illustrations of select specimens, preceded by descriptive tables. Half the volumes also include prefatory material. As the complete collection runs to nearly 700 pages, it was possible here to include only fragments of the whole. Since all Ruysch's illustrations have been reprinted in this book, it was a given that I would translate all the explanatory tables. I have also translated all the introductions.[47] While this entailed including fewer medical specimens, my defense is that these descriptions are generally quite dense and often caught up in archaic medical controversies that will baffle anyone without expert knowledge. On the other hand, Ruysch's introductions provide fascinating personal insight and historical context that will enrich the experience of any reader. To give the feel of Ruysch's unabridged *Thesaurus,* however, I translated roughly the first half of the first book without any cuts.

As for the minutiae of my translation, I have only a few points to make.

Latin favors long sentences with fluid word order. These are features I tried to retain, without unduly straining modern English. I also tried to convey a period flavor, but with a light hand: *'Twas*es and *'twere*s I have mostly confined to the contemporary poems that Ruysch included, which were self-consciously antiquarian even at the time of their writing (and which appear almost cringingly fawning to a modern eye). While I did consult a seventeenth-century Latin-English lexicon, I made no special effort to avoid anachronisms. The word *publicize*, for instance, may not have been coined until 1844, but it was clearly a concept that Ruysch understood.

A weightier choice was my decision to use ordinary English words instead of medical terms when possible. This wasn't a decision I made lightly, because

Ruysch's text is of course chock-full of medical language, much of which would be familiar to a modern audience—words such as *tibia*, *femur*, and *intestine* for instance. By preferring *shin*, *thigh*, and *gut*, however, I hoped to underscore a linguistic distinction between Ruysch's era and our own. Today, professionals may speak to each other in oftentimes abstruse specialist jargon, but before roughly the mid-eighteenth century, they spoke literally in their own language—Latin. In Ruysch's time, *uterus* was not a more technical or formal word for *womb*; to the un-Latined it was simply foreign.

Finally, there is the question of the name Ruysch gave to his catalogue. The most literal translation of *Thesaurus anatomicus* would be *An Anatomical Treasury*. In my view, *Anatomical Cabinets* would be more accurate, evoking the cabinets of curiosities—those hodgepodge collections of artworks, artifacts, and natural wonders so popular among the cultural elite of the sixteenth to eighteenth centuries. Lying on the far side of scientific modernity, they reflect a worldview very different from ours. The animating principle of these collections was analogy: things that looked alike must be similar in substance as well. This may seem whimsical to us, or at best poetic. But it has its own logic: all that exists is the product of the mind of God, and it bears the stamp of its divine origin. Alignments between objects, then, were more than coincidental; they bound the cosmos in a web of correspondences that inevitably led back to a divine origin.

Ruysch, on the other hand, was living through the peak of the scientific revolution, at a time when the cabinet of curiosity was itself on the way to becoming a curiosity. Although Ruysch had the meticulous eye and skeptical mind of a natural philosopher, in his idiosyncratic curation we can still catch traces of the curiosity cabinets. Not only does he ignore his age's obsession with taxonomy, opting instead to organize his collection according to the principle of whatever "would most charm the eye,"[48] but he also includes a number of objects with no particular medical value but exceptionally high cool quotients: an elephant's tooth with a pistol shot embedded in it; a narwhal embryo; a hen's egg that looks remarkably like a child's penis; and, of course, the fanciful scenes culled from human body parts.

Ruysch's outlook, then, was scientific, but it was not disenchanted. And here we might at last find a thread that unites the *Thesaurus anatomicus*. The God Ruysch worshiped was a craftsman and a connoisseur who delighted in the puzzles presented by obscure patterns and hidden symmetries. All elements of nature—no matter how bizarre—were meaningful, for they still carried a divine stamp, still had a function in God's blueprint. For Ruysch, the scientist and artist, everything points back to God—if only we can teach ourselves to see how it all fits.

Frederik Ruysch
Professor of Anatomy & Botany

The First Anatomical Cabinet
With Engravings

Amsterdam
Jansson-Waesberg
1721

TO THE READER.

THIRTY-SIX YEARS HAVE NOW PASSED SINCE I SET OUT TO SURVEY THE Great Works of Anatomy. But so much of what I encountered in those pages struck me as obscure, nor was it illustrated with sufficient care, that I took up the Dissection[49] knife myself, and to its blade I subjected diverse cadavers, not just human, but those of beasts as well—sparing in particular neither horses nor cows, so that the fine details should meet my gaze with all the greater clarity.

Throughout this time, I would come across various fleshy or softer parts (such as the brain, &c.) that seemed worth preserving, but mastering the art of preservation utterly exhausted my intellect. Just now it would seem that I had discovered a commendable method, but soon after my hopes would be dashed: Then afresh, I would approach my goal by a new path—different, but no less deceptive than before. Thus I tortured my wits for many years, until at length I found a certain technique that I believe will open the possibility of perfectly preserving not just flesh, but even the Brain, and indeed

Icon on previous page (used throughout the rest of the book) is an original typographic ornament drawn from Frederik Ruysch's Thesaurus Anatomicus.

entire bodies, complete with all their internal organs, for centuries, & perhaps forever.

I have in my possession child cadavers, embalmed twenty years ago, that are so neatly preserved that their little bodies seem to be asleep rather than dead.

Now from that time, when I turned my efforts to the dissection of cadavers, it was not at all difficult for me to preserve various vessels, such as Arteries, Veins, Lymph-ducts, Nerves,[50] and also other parts less subject to decay. And so, in the course of time, the harvest from this activity grew to such bulk that a whole room could hardly contain it; & thus I was compelled to acquire a second & then a third room to store my Anatomical collection.

Yet, because it would have been wearisome[51] to do nothing but handle Cadavers, I began to put my mind also to the study of Botany, while still a young man, for a restorative change of pace. With no small pains, I collected & preserved plant specimens,[52] both native and exotic, in such a great number that I needed another small room just for them. Yet not even this was enough to satisfy my intellectual hunger, wherefore I was compelled to nourish it further by closely studying exotic Animals, Fish, Insects, &c.—specimens of which I preserved either by drying or by soaking in Embalming fluid, and which soon entirely filled yet another room.[53]

Having accomplished all this (through God's grace), and feeling the weight of my years, I now have a mind to follow a different path and seek a quieter life; and so I give thanks to Almighty God[54] for the high favor he bestowed in granting me time enough to achieve all that I have been able to do.

In the meanwhile, with the aim of satisfying no few requests, I shall with great pleasure set about making a Catalogue of everything, down to the smallest detail, that is to be found in my Museum, which I shall submit to the judgment of the public, should it please Almighty God to allow me so long a life.

While it would be but a small matter to publish the entire Catalogue all at once, I have nevertheless concluded that it would be better presented in parts, beginning with the items kept in the ground-floor chamber.

In that First Chamber, two India-wood[55] Cabinets present themselves for inspection; and so, in the Second, do four large Cabinets, as well as several small & medium-size Cases, of which:

The First is filled with human hearts.

The Second is full of selected Lung parts, as well as complete Lungs & Hearts, just as observed in their natural state.

The Third contains the reproductive organs of a woman.

The Fourth: The very finely preserved Entrails of several Children.

The Fifth: A throng of many Mouse skeletons, including the skeleton of a leaping Dormouse: these skeletons are all glistening and most elegantly prepared.

The Sixth: Uterine Placentas, & similar things. Also in here are 4 large Chests made of the same India-wood, containing the gleaming skeletons of 4 Youths.

The Third Chamber houses a cenotaph furnished with the skeletons of s everal Children, & also eight large Cabinets, in which are stored various Anatomical specimens, and not a few infant cadavers very neatly preserved for many years.

To begin with, then, I have decided to publish a description of the items kept in the first Cabinet of the first room, with the intention of recounting in succession each of my Objects & everything there is to be noted about them.[56]

Whatever strikes me as unusual or as less easily described, I will illustrate with figures, & in this way each Cabinet will enjoy its own illustrated Catalogue.

That is where the real profit of my enterprise will reside. For no one has ever been able to inspect all my rarities, on account of both the abundance of my collection & the scarcity of time: Even for me, it is impossible to commit to memory every detail—such as which item goes where, which is not an uncommon annoyance.

Once completed, though, this Catalogue will allow me to find exactly where this or that item is kept; so, if anyone at all should read this Catalogue & find something in it that he would like to see, he could be satisfied in an instant. For, while some visitors may be enticed to see one object, another might better please others; & thus I may endeavor to satisfy everyone, without spending undue time.

Moreover, if anyone should doubt that these things really exist in the way I have depicted or described, he will have the opportunity to view them; which is an offer that would not be made by other Anatomists—no small number of whom have attempted to foist on us countless Chimeras that do not merely stretch the truth, but that have in nowise been seen in the human body.

Farewell, then, Kind Reader, & may my labors, which have been completed only at the greatest expense, find favor with you.

The First Anatomical Cabinet.

Elegantly constructed of painted India-wood, it houses the following.

[Item no. I] First of all we come upon a Walnut-wood Pediment, on the bottom left side of the Cabinet, upon which rests an artificial Cliff-scape composed of human stones. The base is natural rock, but it is covered with so many human stones that whatever meets the eye is purely human stone. [Table 1; see image on facing page.]

These were selected from various areas of the body and are diverse in color, shape, size & substance. Among these are:

A. A. A. *Tiny human stones, excreted from the urinary passages, which cover the natural Rock, particularly where the larger stones are not well able to conceal it.*

B. *A human stone in the shape of a walking-stick handle, which we cut out of a Bladder,* post mortem.

C. *A Stone that resembles a Black Mulberry, extracted by Lithotomy.*[57]

D. D. *Stones cut out of the urinary Bladder of an Octogenarian woman, whose remarkable story is recorded in my* Anatomico-Surgical Observations, *published in 1691. (See: "The First Observation.") The patient had suffered for twenty years from an entirely prolapsed womb,*[58] *along with a descending urinary Bladder; accordingly, we took up the whole of the Womb along with the adjacent bladder into our hands, & after sounding with our fingertips, we were convinced that stones were lurking in the prolapsed parts. Nor were we wrong: For upon making an incision there, we extracted within a minute forty-two stones, whose rough edges gave a remarkable resemblance to natural rock. Details of the novel technique by which she was easily cured can be found in the aforementioned Observation.*

E. *A stone extracted from a Urinary Bladder by Lithotomy.*

F. *Another, removed by the same operation.*

G. *A black stone in the shape of a Ginger root, cut from a Renal Pelvis,*[59] post mortem.

H. *Another from a Renal Pelvis, covered in a glistening black, which is commonly seen in stones obtained from this location.*

I. *Another stone, coughed out of the throat after the patient had complained for several years of difficulty swallowing.*

K. K. *Two stones, coughed out of the Lungs.*

L. L. *Two extraordinary stones that closely counterfeit natural Rocks, which were*

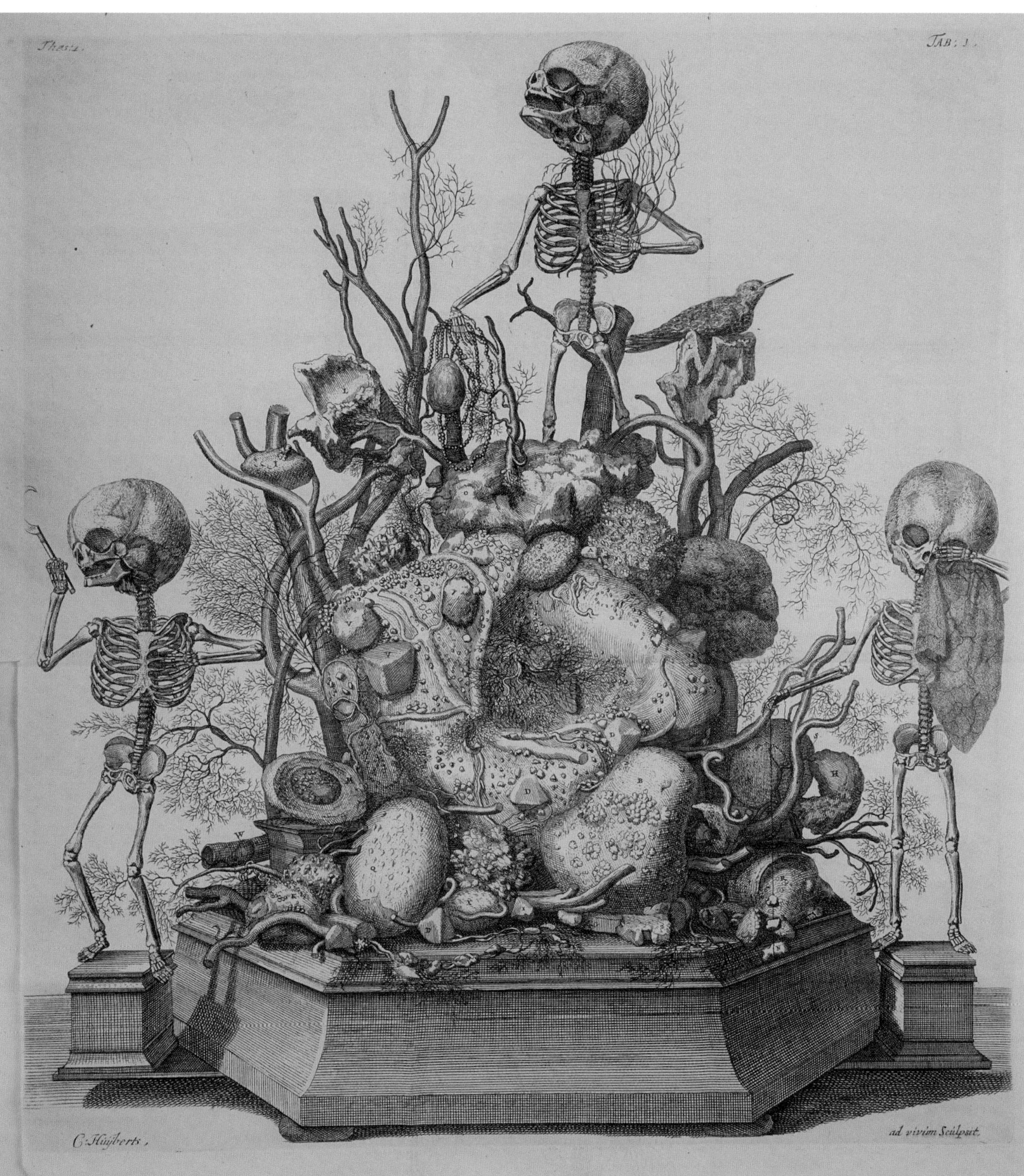
Thesis.
TAB: 1.
C: Huijberts,
ad vivum Sculpsit.

cut from the Breast of an elderly Woman, post mortem.

M. *A dark stone, removed by Lithotomy and covered with a thick, hard, gray husk.*

N. *Tiny stones, taken from the toe joints of a woman with the gout; these hang from a thread held in the right hand of the skeleton of a human fetus about four months old, standing next to the rock landscape. In its left hand, this skeleton clutches a handkerchief, made of the thinnest membrane & embellished with thousands of arterioles colored the deepest red, which it holds to its eyes as if weeping for the woes of mankind. Looking toward Heaven, it is accompanied by this caption:*

Man that is born of woman is of a few days and full of troubles.[60]

O. *A stone of such prodigious size that it rivals a natural Rock, discovered between the legs of a corpse laid out in its coffin.*

P. *One cut from a Bladder.*

Q. *Another, larger one.*

R. *A big stone, in the shape of a Ginger root, formerly attached to a Renal Pelvis.*

S. *A large stone with a gray husk, removed by Lithotomy.*

T. *A very coarse, coal-black stone.*

V. *A few blackish stones, cut out of the kidneys after death.*

W. *A stone that lodged within and ultimately blocked the Ureter of a four-year-old girl, thus trading her life for death. She endured so many torments before returning her soul to God that my heart is moved whenever I recall her memory.*

X. X. X. *Stones removed from a gall Bladder,* post mortem.

All the rest of the stones were either passed through the Urinary passages or removed, post mortem, from the Bladder or the Kidneys.

Interspersed among all this are arterial branches, the blood of which has been artfully replaced with a ruddy-colored wax preparation, and which represent trees with bare branches. These help hold together the covering of little stones and provide greater stability to the overall Cliff scene while also charming the eye. The said arterial branches have been selected from various body parts of both men and beasts.

About this Cliff stand the Skeletons of three miscarried[61] fetuses, each about four months old. I have already made mention of the first one, above.

The second, placed on the opposite side, wields a sickle in its right hand, as though threatening to strike, and it bears this caption:

Death spares not harmless youth on bended knee.[62]

The third skeleton, of about the same age, is placed atop the highest part of the Cliff, bearing in its right hand a thin strand of immaculate pearls, at which it appears to point with the forefinger of the left hand. Head upturned, as though looking toward heaven, it bears this caption:

Why should I treasure the things of this World? [63]

[Item no. II][64] Next to the Cliff scene is a second walnut-wood Pedestal, which is furnished with Wind-pipes[65] from both people and calves. In the center of this pedestal is another walnut pedestal, occupied by the skeletons of two fetuses, miscarried at about five months; one of which in its gestures appears to be mocking human affairs, & thus represents Democritus;[66] while the other, holding a kerchief to its right eye, represents Heraclitus.[67]

The caption for the first skeleton reads thus:

To live is to endure great ills; while I,
By death released, rejoice in silent triumph.[68]

And for the Second:

We babes, of sweet life robbed and seized from mother's breast,
Were carried off one dismal day and given to untimely death.[69]

Note 1: The said handkerchief, made of membrane, enjoys an embroidery of countless red arterioles, whose serpentine course presents a not inapt comparison to the divine handiwork that the Prophet David speaks of in Psalm 139.[70]

2. Here it is plain to see that the bones of the Cranium grow in striations, like ice in winter.

The lower Pedestal, mentioned above, contains:

A. A Calf's wind-pipe, filled with a red waxy preparation and turned upside-down, so as to resemble a tree with many leafless branches.

Note: I have dried the branches in such a way that they retain a natural disposition: that is, with some bending forward and others backward.

B. A sample of a calf's Lung, inflated, dried, & and sliced into layers, to expose to view the vascular, or spongy, substance of the Lung.

C. A sample of human Lung, prepared as above, so that not only the vascular substance may be observed but also stones that were formed within the Lung.

D. An infant's Wind-pipe, inverted, with the pulmonary Arteries and Veins still attached.

E. An inverted Wind-pipe, from a fetus of some four months.

F. Part of a Wind-pipe, filled with whitish wax, across which run the branches of the Bronchial Artery. Taken from the body of a calf.

G. The Wind-pipe of a calf, prepared according to an older technique, & preserved now for thirty-four years.

Item No. III. Between the two aforesaid miniature Landscapes[71] is the snow-white Skeleton of a Girl of three or four years, seated on another wooden Pedestal, & prepared according to my own technique, by means of which the bones are connected almost entirely by their natural & particular ligaments, and also the arteries of those ligaments are filled with a waxy material, whence these joints receive an infusion of the deepest red coloring. Examining these said joints under a magnifying glass would reveal not just the twig-like capillaries but many parts even more delicate.

Note 1. The outermost part of the Head bone,[72] properly styled "the Pericranium," boasts so many wax-filled Arterioles that this aforesaid pericranium is suffused with the deepest red.

2. The left hand bears a Branch of Splenic Artery, taken from a calf's Spleen and also filled with wax.

3. The right hand holds a silken thread from which hangs a Boy's Heart, both ventricles of which, along with their connecting Auricles, have been filled with a waxy substance and rendered hard as stone.

4. The aforesaid Heart includes the trunk of the Great Artery[73] exactly the way it emerges from the left ventricle of the Heart, retaining its natural diameter, shape, & arc.

5. Here it can also be seen how the three branches (namely, the two Carotid Arteries & also the left subclavian Artery) extend at a steep angle from the upper part of the Great Artery. That the right subclavian Artery has its origin in the right Carotid Artery is also clearly apparent.

6. Furthermore, one may observe that both the Vertebral Artery (A.A.) & the internal Mammary Arteries[74] (B.B.) issue from the subclavian Arteries (C.C.).

7. One can also discern that four branches emerge from a membranous sac (or the base of the Pulmonary Artery): I have illustrated this in my collected *Anatomical Letters.* (See my response to the Tenth Problem.)

8. Similarly, the naked eye can also clearly see the way the Pulmonary Artery divides, first to the right & the left on either side the Heart, and then it further subdivides.

9. The Coronary Vessels sown across the surface of the Heart are a delightful spectacle to behold, while still tinier capillaries may be easily seen under a magnifying glass.

This Heart, as I have said above, hangs from a silken string, and bears this caption:

All Human matters hang from a slender thread.[75]

Item No. IV. To the right of the aforesaid Skeleton is a jar that contains the head of an infant, so well prepared & preserved, and with such lifelike color (though without any cosmetics) that it appears to be asleep rather than awash in embalming fluid, even though it has been preserved for many years.

Note 1. The top of the Cranium has been removed, and the Brain extracted, in order that the ten pairs of nerves[76] that pass through the bones of the Cranium, not excepting the Pathetic nerves or the Accessory nerves, might be observed more clearly and distinctly, than in a recently deceased head.

2. It can also most clearly be observed that the hard branch of the auditory Nerves[77] is distinct from the soft branch, particularly on the left side.

3. I have opened the lateral folds of the dura Mater[78] to bring the four folds into view.

4. The Stalk of the Pituitary Gland is suffused with the deepest red on account of its abundance of red-wax-filled arterioles.

5. The Pituitary Gland, sitting deep in the Saddle,[79] protrudes somewhat in the middle, where it connects to the Stalk.

6. To the side of this Gland, & also to the side of the Optic Nerve, the severed trunks of the Carotid Arteries can be discerned.

7. The features of the face, & indeed even the lips, are not at all shriveled, & are endowed with lifelike color.

Item No. V. To the left of the skeleton discussed above is another jar that contains an Infant's head, floating in the same liquid and prepared in the same way as above.

Note 1. At the base of the interior of the Head, to the right of the pair of vagus nerves, is found a supernumerary nerve.[80]

2. The extremities of the Olfactory Nerves can be clearly seen to pass through sieve-like openings in the Bone.

3. In this head, eleven or twelve pairs of Nerves can be seen passing out of the Cranium.

4. On each side, the vagus nerves pass through the Cranium in two places, with a little piece of bone between them.[81]

5. Here it can also be seen that both the right and the left parts of the accessory Nerve consist of various sinewy fibers that emerge from various locations.

6. This head also shows how tendinous the Dura Mater is.

Item No. VI. Behind the heads just now described, the small guts of a Boy, aged 3 or 4 years, present themselves to view; dried and held together by their original Mesentery,[82] they reveal their natural circumvolutions. I have filled the guts to capacity with wax so that they should regain the taught rotundity with which food, drink, and air endowed them in the living body. A remnant of the Colon is still attached, as well as the whole Blind Gut with its worm-like Appendix, so that their composition at that age can be seen.

The following items are found in the upper Shelves or Repositories of the said Cabinet.

The First Shelf, or Repository.

Item No. I. The small guts of a newborn human infant, stuffed and dried, along with the connecting part of the Colon, as well as the Blind Gut and the Appendix, all prepared for the same end—that is, to approximate the structure of a typical Blind Gut at this age, which is far different from what we see in children or in those of more advanced age; for in newborns, the diameter of the Blind Gut tapers quite gradually, without noticeable swelling, & thus it dwindles to the Appendix; however, in adults, it in fact swells suddenly from one point, & from that protuberance the Appendix emerges; and so, at that age, the diameter of the Blind Gut diminishes in a manner far from gradual.

The natural capacity of these guts has been preserved, & they are stretched taught and without wrinkles.

Item No. II. A glass jar of embalming fluid in which floats a sample of Splenic Artery from a calf, filled with a red waxy substance instead of blood, whose extremities resemble worn-out brush tips, just as I have recorded, in a human example, in Fig. 4 Table 4 of my Letter in Response to the most Expert Master Campdomerc.[83]

Item No. III. [Table 2; see image on following page.] An infant's Forearm in embalming fluid, endowed with a natural complexion so glowing and fair (without admixture of any red coloring) that it has scarcely lost its living disposition. Its hand holds an artificial polyp,[84] made from hog's blood in the following manner: Blood from a freshly slaughtered hog was shaken vigorously in the hand until cool, by which point it had produced in abundance a substance resembling polyp tissue; in the substance produced from a single agitation, such as I have described, various particles of fat appeared that cohered into a solid, polyp-like mass (as represented in Figure, Table 2, E). With assiduous shaking, this aforesaid fat, rather firm and quite white, assumed the form of globules of various size, covered by a shell-like membrane, and, with further force, this combined into the particular polyp discussed above; meanwhile, the hand that had agitated the blood was pristine, befouled with neither grease nor cruel gore.

These facts suggest that blood itself contains dissolved fat, which is deposited into cells when it congeals.

It should also be noted that this jar containing the Forearm, & all the other jars as well, have been sealed not with bladder lining, as is usual, but with covers selected from various body parts, such as gut, human skin, membranes

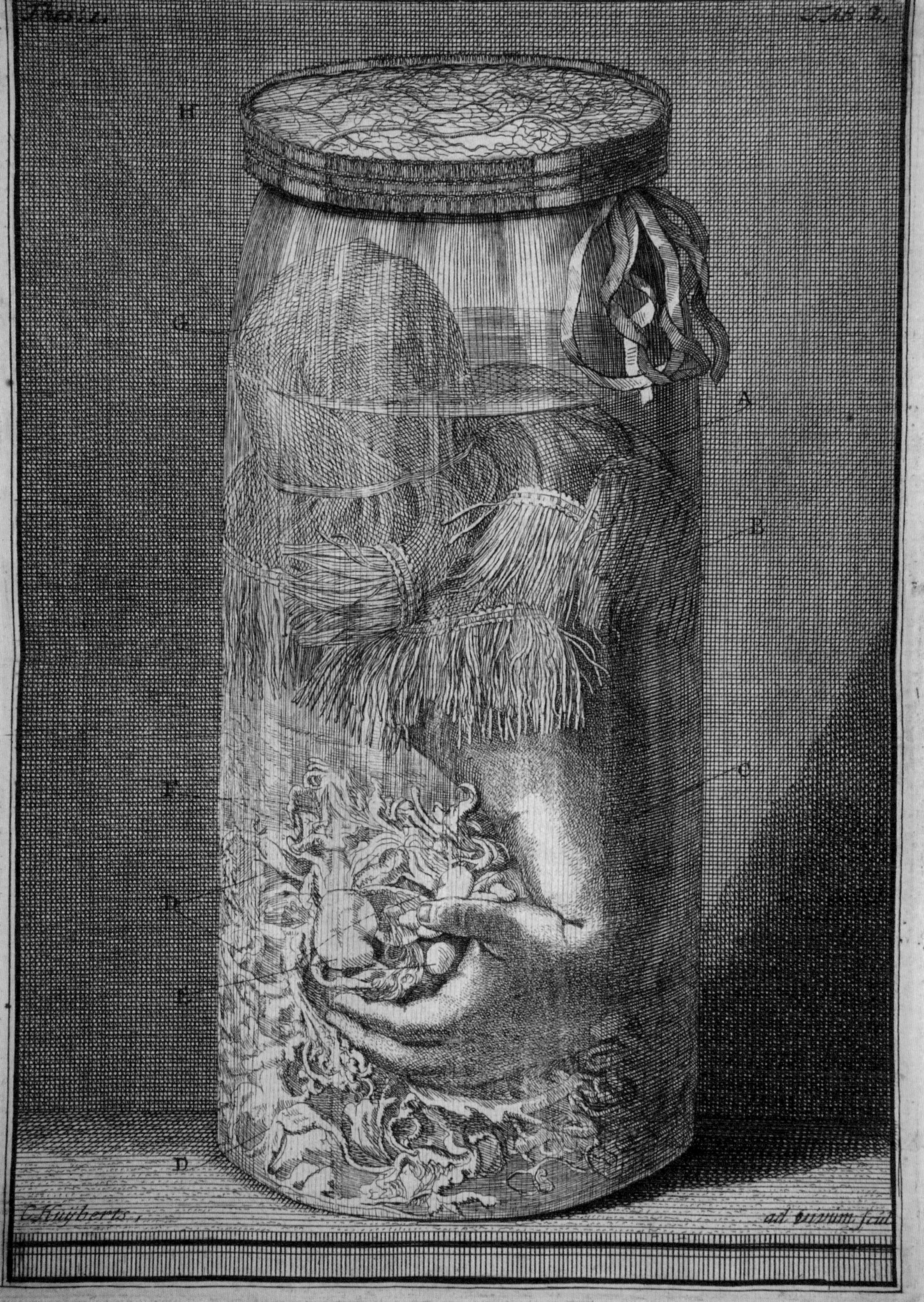
Thes. 2.
Tab. 2.
H
G
A
B
C
F
D
E
D
C. Huyberts,
ad vivum fecit

of the Brain, &c, all of them teeming with blood vessels that have been filled with red wax, not merely to please the eye, but also to make the crawl[85] of the various vessels inside them more easy to discern.

[...]

TABLE 3 EXPLANATION [see image on page 125][86]

Fig. 1. Shows a cluster of Teeth, taken from a female Sheep.

A. *A Canine Tooth.*

B. *A Molar.*

C. *Part of the cluster, in which other Teeth are still enclosed, without any of them exposed at all.*

D. *A piece of the thick membrane, in which the Teeth are partly covered, and partly exposed.*

Fig. 2. Represents the same Teeth, seen from another side, with a small piece of the said thick membrane.

A. *A Canine Tooth.*

B. *A Molar.*

C. D. *The lower parts, or roots, undeveloped, neither hollow, nor solid & bony, but rather endowed with a stony substance.*

Fig. 3. Displays a piece of the breast—to wit, several ribs, & the intercostal Muscles, with their exterior intercostal Arteries; this is from an infant.

A. *Severed ribs.*

B. *Intercostal Muscles.*

C. *Intercostal Arteries, exterior, as I said, but running not only throughout the upper edge of the ribs, but the lower as well. These trace their origin, not directly from the trunk of the great Artery, and not from the internal mammary Arteries either, but, in fact, the interior intercostal branches issue forth these said exterior arteries.*

Fig. 4. Shows a boy's Tongue, seen from beneath.

A. *The membranous covering of the Tongue, looking rather thick on account of its inversion.*

B.B. *The Sublingual Arteries, whose branches are most numerous, resemble a little forest of tiny, leafless saplings.*

Fig. 5. Represents Flies & Nymph casings.

A. *A tiny Fly, still sticking to the casing, from which the fly has emerged.*

B. *The same, as seen & drawn through a microscope.*

C. *Two casings sticking to each other, from which the flies have emerged.*

D. *A casing, seen on its own.*

TABLE 4 EXPLANATION [see image on page 126]

[Fig. 1]

A. *A piece of breast skin.*

B. *The Nipple of the breast.*

C. *The Papillary nerves*[87] *of the Nipple, projecting beyond the true skin*[88] *& c learly visible to the naked eye, but exaggerated somewhat in this representation.*

D. *The Cavity in the tip of the Nipple.*

Fig. 2. Shows a Whale's common milk-bearing Duct, into which all the other ducts converge and deposit their milk; this duct is furrowed with many wrinkles.

Fig. 3. A sample of the Pia mater[89] of a newborn fetus is seen here, which is so slender and delicate at this age that it nearly surpasses the slenderness and delicacy of a spider's Web.

It should be noted here that the blood vessels are dispersed throughout the said membrane in such abundance that the blood Vessels seem to contribute more to the continued existence of this Pia mater than the remaining membrane does, inasmuch as the said vessels hold in place the overall design, and are not jumbled together.

Fig. 4. Displays the Nipple of a woman's breast, with the areola attached. A. *Part of the Areola, without the Epidermis, which is why the Areola has whitened.*[90]

B. *Part of the Areola, whose Epidermis has not been removed, hence it appears darker.*

C. *The aperture, or mouth, of the milk-bearing duct visible in the Nipple of the breast.*

D. *The Papillary nerves project more than a little in the nipple, as well as the Areola, & are more visible, under a microscope, here than elsewhere in the skin; whence Nipples, deprived of the Epidermis, incite terrible anguish, certainly more than in other parts of the body, where the Papillary nerves push out less far beyond the true skin. Hence it is that cuts, scrapes, &c. in the Nipple are more difficult to treat. On the other hand, in the case of a well-constituted Nipple, with the Epidermis intact, a certain great itch & pleasure is aroused in nurses by suckling infants, on account of the repeated motion & pressing of the Nipple during the time of sucking; but not so, however, if the infants, grabbing at night, should take hold of much more than the mammary Nipple.*

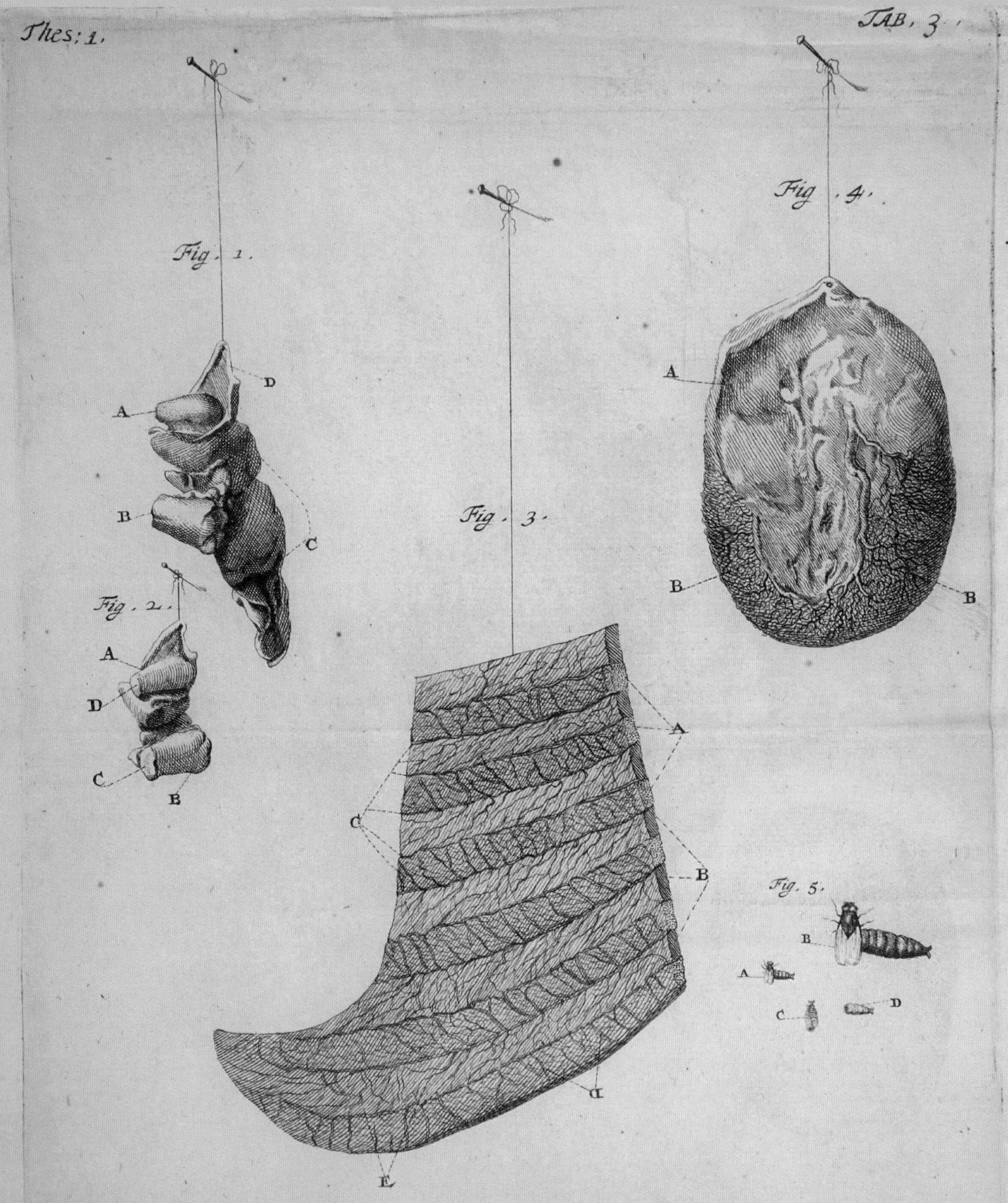
Thes: 1.
TAB, 3.
Fig. 1.
A
B
C
D
Fig. 2.
A
B
C
D
Fig. 3.
A
B
C
D
E
Fig. 4.
A
B
B
Fig. 5.
A
B
C
D
C. Huijberts, ad vivum sculp:

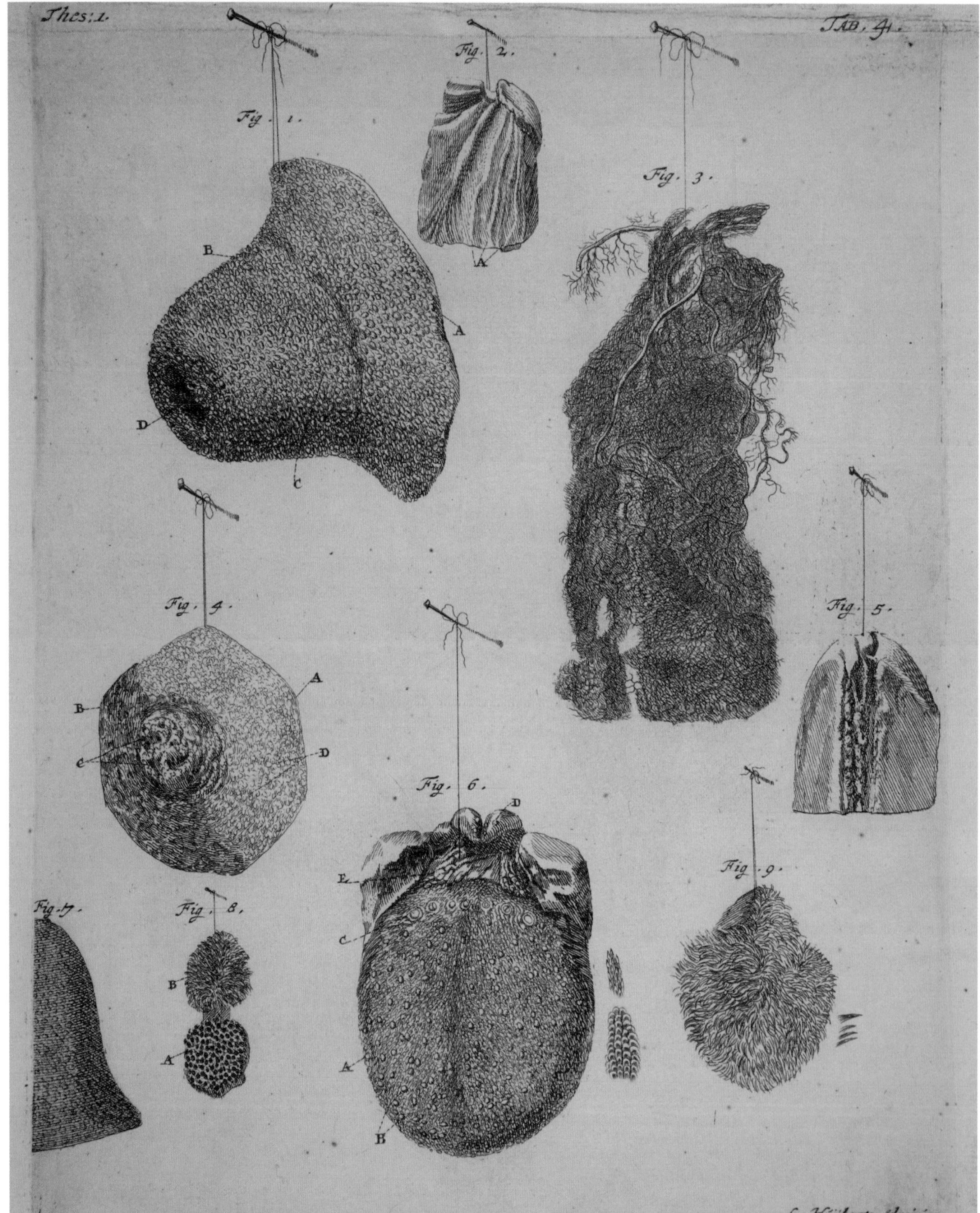
Thes: 1.
TAB. 41.
Fig. 1.
Fig. 2.
Fig. 3.
Fig. 4.
Fig. 5.
Fig. 6.
Fig. 7.
Fig. 8.
Fig. 9.
A
B
C
D
E
C. Huijberts, ad vivum. sculpsit.

Fig. 5. Outlines a piece of Whale's Nipple, whose milk-bearing Duct is furrowed & fringed.

Fig. 6. Presents the image of a Tongue, with the Epiglottis, across whose surface are dispersed innumerable Buds,[91] which are something very like an organ of taste.

A. The Tongue.

B. Buds in the shape of tiny Mushrooms, among which are seen dispersed countless arched Buds. I took care to sketch these buds larger than actual size, as seen under a microscope, to the side of this Tongue, & below those that grow out into points, the Image of which I also added, seeing that I have very often encountered these tapered buds.

C. A row of large Buds which are flat and round, with minute apertures, or mouths, each circled by a ring.

D. The Epiglottis, covered with various Glands.[92]

E. Large Buds planted at the root of the Tongue, below the Epiglottis.

Fig. 7. Presents a piece of Whale breast, whose Epidermis, & Malpighi's Corpus reticulare,[93] have not yet been removed, which is why the papillae of the skin are not pushing out further.

Fig. 8. Offers a small piece of Whale's Nipple, with its Papillary nerves separated, and from which hangs a piece of corpus reticulare, perforated with innumerable apertures.

A. Corpus reticulare.

B. The separated Papillary nerves.

Fig. 9. Represents a piece of Whale Nipple, whose Papillary nerves I have loosened in plain water; the individual Papillae are clearly constituted of many nerve fibers, but, in the manner of a wet paint brush, the fibers here diverge from each other.

It should be noted that in loosening the nerve I applied no force: for I removed only the layer with which the nerve was covered. This experiment seemed to confirm that the Papillary nerves consist of the tips of ordinary Nerves, inasmuch as the Nerves are also composed of innumerable fibers. In a person, however, I could never produce this loosening of the Papillary nerves, inasmuch as those fibers adhere firmly to one another.

To the left side of the said figure are found a few Papillary nerves that have not been loosened, which end in sharp points.

END.

Frederik Ruysch
Professor of Anatomy & Botany

The Second Anatomical Cabinet
With Engravings

Amsterdam

Jansson-Waesberg

1722

TO THE KIND READER.

IN THIS, THE SECOND CABINET, & IN OTHERS TO FOLLOW, YOU WILL find, kind Reader, some few objects that I have already mentioned in the first. I would have you know, however, that I have not done this without reason. For it was only after I published the description of my first Cabinet that I at last grasped various details in my preparations that eluded me while writing. These observations are refined again and again, because I persist in dissecting & examining cadavers without cease. Likewise, you will notice that the preparations are not uniform, in terms of how each is placed next to this or that other object. At the time I submitted the description of my first Cabinet to the judgment of the public, a method of treating the liver without recourse to the so-called glands (which, in fact, do not exist)[94] still eluded me, as did a technique of preparation by which all the vessels, which constitute nearly the whole liver, could be brought into view, including the tiniest capillaries. However much I have now learned about these & many other questions through experimentation, I nevertheless cannot bring myself to throw away my earlier objects, & replace them with new ones. Now, let anyone examine this or inspect that of what I have prepared over 37 years: & it will be clear that there is an order to them. Having assembled these specimens with infinite labor, I preserved them in 16 & more Cabinets and Chests, into which they were placed, not sorted by type, but in whatever order I thought would most charm the eye; it would have been easier for me to fill one Cabinet solely with human hearts, prepared by various methods, and another with embryos and fetuses, and yet another with children's heads; but that would have been less pleasing to behold.

The Second Anatomical Cabinet.

In this second Cabinet, six Shelves, or repositories, are found, on which, from top to bottom, one may inspect the following.

[...]

THE THIRD SHELF

[...]

ITEM NO. VI.[95] A HYDROCEPHALIC FETUS, SEVEN MONTHS OLD, OR rather between six and seven months, whose massive head dwarfs both its torso & limbs, and about which stand five Jars, containing something quite unheard of: wherefore I think that I should briefly recount its story. A certain honest matron, still alive today, & living in good health in Amsterdam, had some few years ago fallen into labor. The attending women bade me be summoned, seeing that the delivery was not proceeding as would be hoped; before I arrived, though, she had already given birth to the aforesaid hydrocephalus. However, the midwife, as she was removing the afterbirth, realized that the womb was still laden, and in my presence she extracted, in several successive attempts, a mass of heterogeneous matter—fibrous, cellular, soft, hard, and spongy[96]—which I have preserved in five separate Jars, and in which the following appear:

The first Jar (labeled as Item No. VII), contains a remarkable sample of the above-mentioned heterogeneous substance, out of which emerges an entire leg, complete with foot, of a two-to-three-month-old fetus, but without any visible body. Near the point where the leg appears out of the said substance, another foot can be discerned breaking forth, but without any visible leg. If the Jar is turned over slightly, then one may again perceive within the said substance the ulna & radius of another fetus of about the same age, from which the flesh has fallen away because it had become exceedingly soft & crumbly from irregular & inexpert handling in the time before the lady's family decided to entrust the specimen to me.

Below the said foot, a single finger emerges apart from any visible hand; I have marked this place with a pin, so that it may quickly come into view. Opposite from this are found two feet with neither shins nor thighs; there are other parts, too, but so very commingled that nothing can be determined with certainty. (See Tab. 4.[97] Fig. 3. Cabinet 2.)

In the second Jar (labeled No. VIII), one can also see a large portion of the said heterogeneous substance, along the top of which runs a thigh bone, bereft of both its flesh & covering of membrane;[98] I have indicated the cartilage of the knee with a pin. Below this knee one can also make out another knee, below which protrudes merely a single toe; and still below all of these, diverse members—such as feet and hands, more or less developed—come into view. (See Tab. 4. Fig. 1, Cabinet 2.)

In the Third Jar (No. IX), a larger mass of the aforementioned substance is preserved, in which one may inspect the following:

From the bottom of this substance sprout two deformed legs, to which an arm is joined, emerging from the same place. The Shins of these limbs are severely bent.

The said limbs originate from a small ball of this substance, that may be likened to a tiny belly, provided with a little pendulous part in the middle, where the penis would be.

If the Jar is turned over, a foot without any visible shin comes into view. (See Tab. 4, Fig. 2.)

The fourth Jar (No. X) holds an exceptional specimen of the above-mentioned substance, from the bottom of which descends a buttock, with an entire leg (that is to say, with Thigh, Shin, & foot), across the surface of which snake the tendons of the muscles, in full sight, for the fleshy matter has been worn away.

The aforesaid foot is misshapen, for the lower joints[99] of the Shin & thigh bones are not in the right place. Whether this stemmed from a malformation during gestation or afterward, I cannot determine.

Above the aforementioned buttock, and lying aslant, is found another shin, with the connecting calf-bone, both of which are bereft of muscular matter, and so may be observed laid bare.

Above the said bones, a foot emerges, also from the aforesaid mass, that appears to have no connection with any of the bones just cited.

Tilting the said Jar brings into view the rudiments of two hands, each of which has three fingers, undeveloped but with nails. (See Table 4, Fig. 5.)

The fifth Jar (labeled No. XI) contains a sample of the same substance, in which no fully developed parts are visible.

Note the following:

1. All of the parts cited just above appear to be about three weeks old.

2. Nowhere is there to be seen a fully developed head, although round bodies are mixed throughout, giving the false appearance of little heads.

3. The bones reported above were dressed in soft flesh, when first brought to light; however, clumsy handling wore the flesh away to nothing in various places, before this was entrusted to me.

4. This subject strikes me as an open field for debate, which is certainly beyond my present scope, since, as previously stated, I am striving for brevity. Just so much will I state: that I am of the opinion that, at the point in time when that honest matron was got with child, several eggs were simultaneously affected, or fecundated: hence their commingling followed, when they passed together through the ovarian ducts and into the womb.

5. Also note that in an earlier marriage that woman was never with child, & that the labor which I have discussed was her fist; this was in her second marriage, during which she later bore several healthy children.

[...]

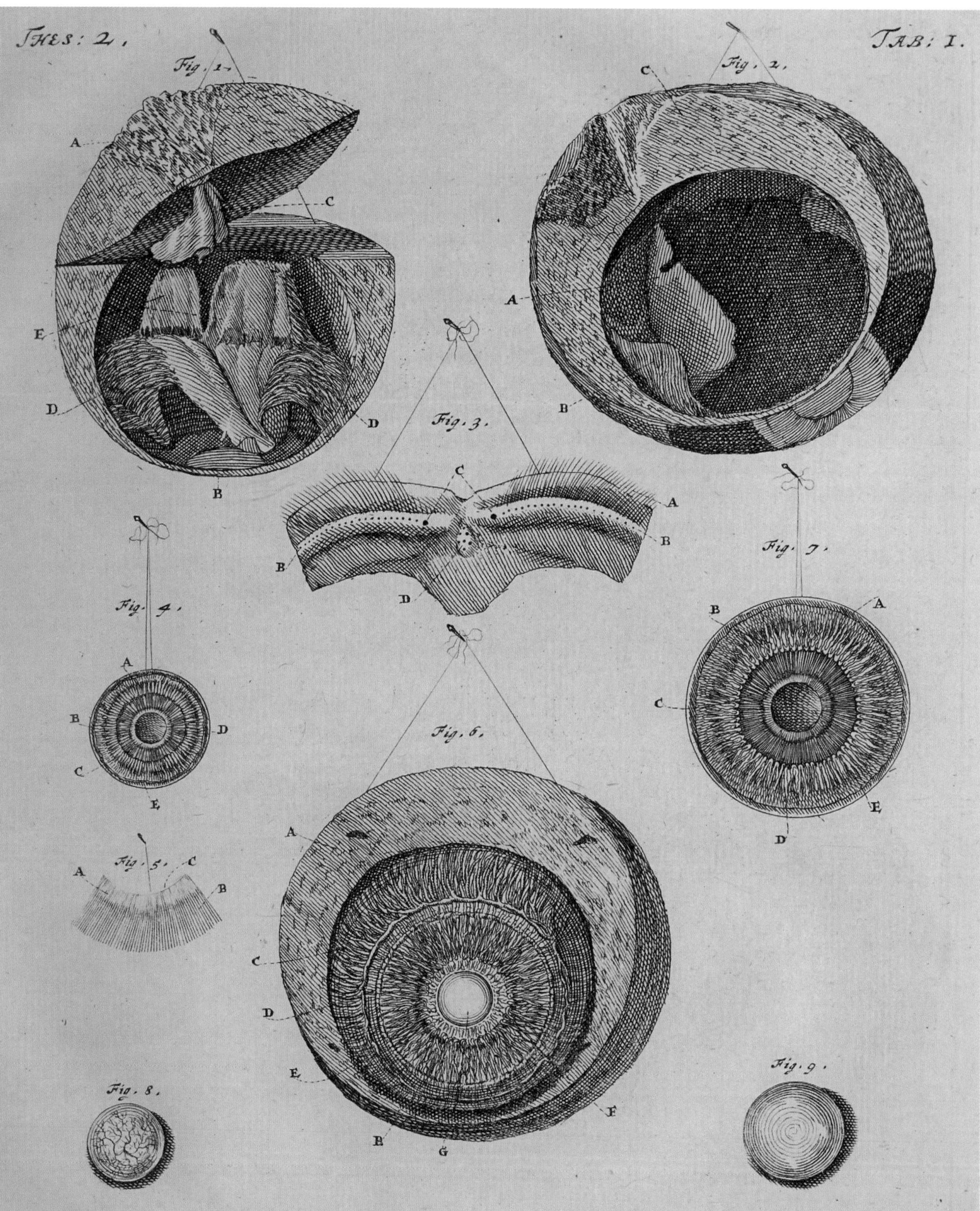

C: Hüyberts. ad. vivum. Sculpsit.

Anatomical Cabinet II

EXPLANATION OF THE FIRST TABLE [see image on facing page]

Figure 1. Shows the eye of a Whale, which has so been dissected that we may see not only the Sclerotic coat,[100] but also the constitution of the Ruyschian[101] & Choroid coats.[102]

A. The fleshiness of the Sclera near the optic nerve.

B. The thinness of the same on the pupil side.

C. Part of the retina.

D.D. Portions of the two choroidea, seen from the exterior, bent backward, & peeled away from the Ruyschian coat. Across this surface is scattered a dense row of blood vessels.

E. Portions of the two choroidea, peeled away from the Ruyschian coat, so that they still adhere to the lower part, whence they seem to have been united with the Ruyschiana. (Note: through this said face, the above-said vessels are less visible than those in the exterior face marked D.D.)

* Denotes the Ruyschian tunic, which in a whale's eye attains the same thickness as a sheet of paper.

Fig. 2. Represents the eye of a whale, bisected lengthwise.

A. The Retina.

B. The Choroidea, joined to the interior surface of the Ruyschiana.

C. The Sclera.

Fig. 3. Displays the better part of twin eyelids from a man.[103]

A. Eyelashes.

B.B. The mouths of the ducts, through which a certain humor is filtered.

C. *Punta Lachymalia.*[104]

D. The tear gland, in which various apertures, or mouths, are seen gaping.

Fig. 4. Represents a human cornea, seen from the back side, in which the margin of the cornea, the *ligamentum ciliare,*[105] the *processus ciliaris,*[106] lesser circle of the Iris,[107] & the pupil can all be seen.

A. *Ligamentum ciliare.*

B. The mucosal *processus ciliaris.*

C. The smaller Circle of the *ligamentum ciliare*, in which the tendons of the *processus ciliaris* end. The said smaller circle, seen from behind, is called by me *the lesser Circle of the Iris*; for the Iris is the outer face, and the *ligamentum ciliare* is the interior face.

D. The pupil.

E. The margin of the cornea.

Fig. 5. Shows the mucosal *processus ciliaris* with its tendinous fibers &c. It should be noted that this illustration was drawn from a microscope, so its natural size has been increased by many orders of magnitude.

A. The tendinous substance of the *processus ciliaris.*

B. The mucosal substance of the same.

C. The muscular fibers of the smaller circle, designed for constricting the pupil, which the engraver has represented here in creditable detail, for they cannot be seen very clearly in the actual object.

Fig. 6. Presents a Whale's eye, bisected crosswise.

A. The Sclera, whose thickness is surprising.

B. The Sclera, whose thickness appears here somewhat diminished, but, since the Choroidea projects out beyond the Sclera, its true thickness cannot be seen.

C. The Choroidea.

D. Blood vessels running across the choroidea.

E. Arteries encircling the entire Choroidea, out of which emerge countless upward-crawling branches.

F. The lesser Circle of the Iris. The major is marked with an asterisk.

G. The Pupil.

Fig. 7. Shows the eye lens of a calf, seen from behind. In this all the major parts look just as in a person.

A. The margin of the Cornea.

B. *Ligamentum ciliare.*

C. *Processus ciliaris.*

D. The lesser circle of the *processus ciliaris*, across which run the tendons of the *processus ciliaris*, in a dense row, & on a direct path.

E. The Pupil.

Fig. 8. Shows the crystalline humor[108] of a calf, with its surrounding membrane, called a "web," across which are scattered many filled arterioles.

Fig. 9. Displays for viewing the crystalline humor from a Whale. The smallness of this, & of the eye too, are quite surprising, considering the bulk of the body.

EXPLANATION OF THE SECOND TABLE [see image on following page]

Fig. 1. Displays a very large piece of a human Bladder, inverted, so that its interior surface is now outside, endowed with countless membranous excrescences & branches, owing to abrasion from stones, with which the patient was greatly troubled.

A. A prodigious thickness of the Bladder, which is common enough in cases of stones.

B.B. Branching & membranous Excrescences.

C. The bottom of the Bladder, without any excrescence, inasmuch as it was not injured by stone.

D. The exterior surface of the Bladder, which, by inversion, has become the interior.

Fig. 2. Shows a sample of Vessels, from the womb of a pregnant woman, whose fibers are covered all over with very tiny grains of sand.

Fig. 3. Shows a sample of liver taken from a patient suffering from the Dropsy.

A. The surface has been made exceedingly uneven from its cellular composition.

B. Cells, or bladders, containing a muddy substance; so numerous, that the entire liver, such as there is, consists of these.

Fig. 4. Displays a small stone, ejected from a horse's stomach.

Fig. 5. Shows a piece of a large stone ejected from the stomach of the same horse, which has been broken in two by sawing and by main strength, and in whose center was found a cavity, formed by an oat seed that had adhered to the bowel, where the oat became covered in a stony substance.

A. Various stony layers, from which the exterior portion of the stone was constituted.

B. The interior substance of the stone, in which the layers are less visible.

C. The pit produced by the oat.

Fig. 6. Displays the other part of the said stone, in which the oat seed remained, after it was broken.

A. The part cut by the saw.

B. The part broken by force.

C. The oat seed stuck in the middle of the stone.

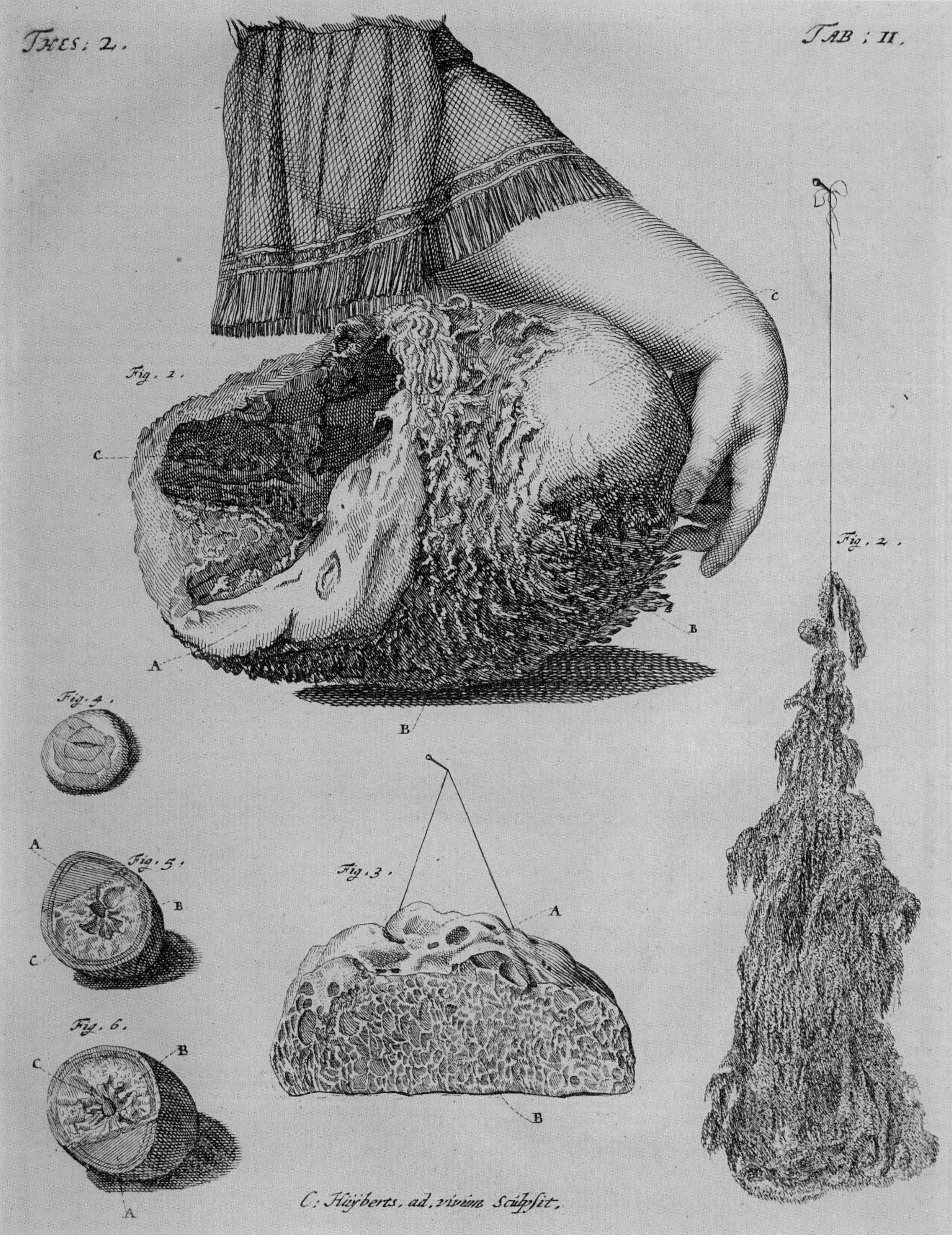
Thes: 2.
Tab: II.
Fig. 1.
C
A
B
C
B
Fig. 2.
Fig. 4.
A
Fig. 5.
B
C
Fig. 3.
A
B
Fig. 6.
C
B
A
C: Huÿberts. ad. vivum Sculpsit.

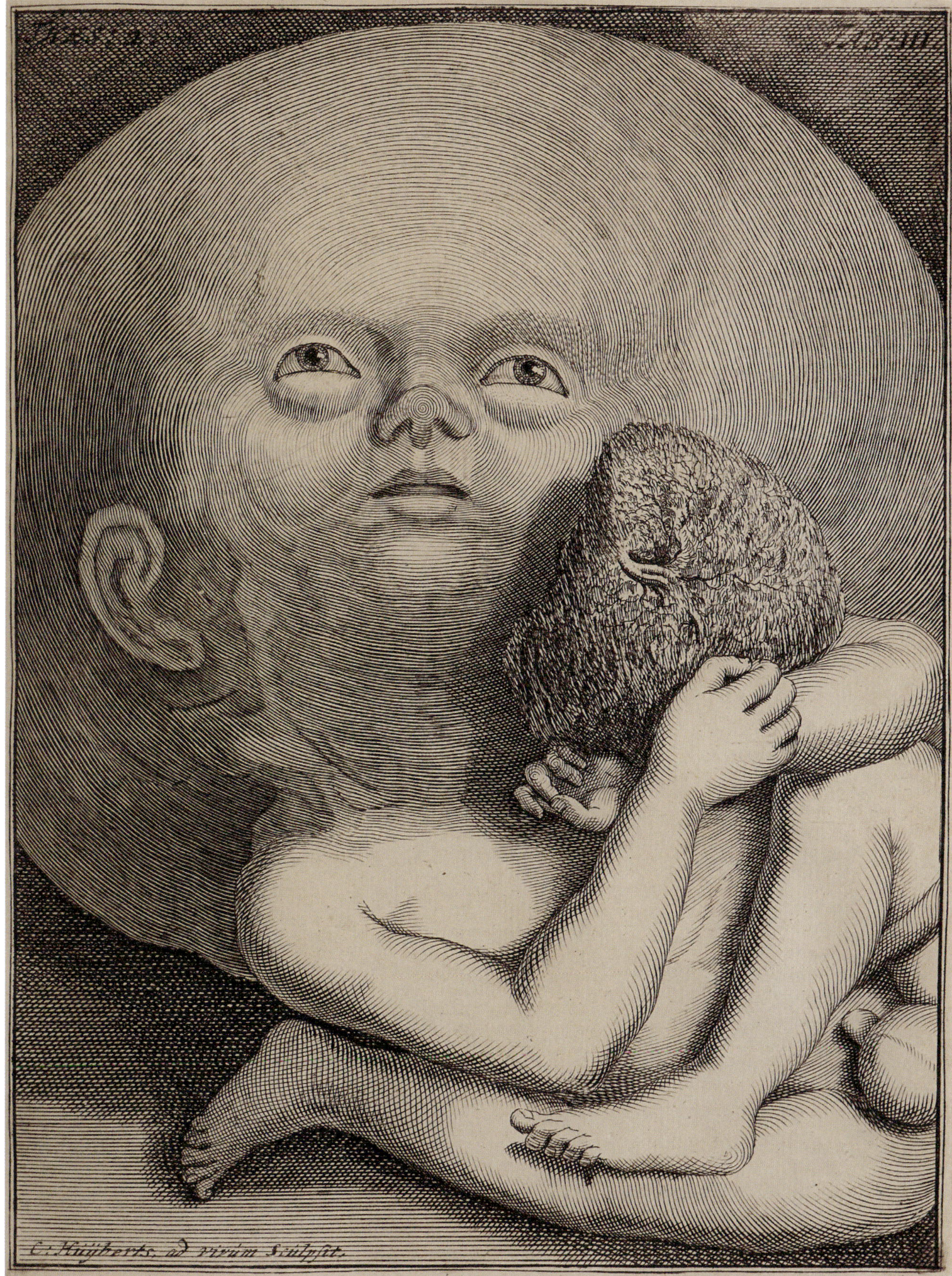
C. Huijberts, ad vivum Sculpsit.

It should be noted that the Noble Master Blok took a piece away with him, so that only one part is found in my Cabinet.

Table III [see image on previous page]

This shows a hydrocephalus of six or seven months, preserved in embalming fluid, whose head is of prodigious size.

Between the hands & head, I have positioned a sample of Placenta, in such a way that the head holds it in place; this said sample is only roughly illustrated, since it was not my intention to display the true constitution of the placenta here; the Kind Reader, however, will find that in Table VI, fig. 2 of this second Cabinet.

Table IV [see image on facing page]

This shows five Figures of a certain heterogeneous substance, copiously laden with the parts of many fetuses, about three months in age, & that were brought to light shortly after the above-mentioned Hydrocephalus, and from the same woman.

Fig. 1. Displays a sample of the heterogeneous substance, laden with the said body parts.

A. A bare thigh bone, stripped of flesh and membrane.

B. A single digit of either a foot or a hand; which one, it is not easy to tell.

C. A cartilaginous kneecap, located near the knee of another thigh bone.

D. The cartilage of the knee.

E. A foot.

F.[109] A hand, whose digits are undeveloped & ill-formed.

G. Another foot.

H. A developed hand.

I. Two less developed digits.

Fig. 2. Presents a large mass of the aforesaid substance, replete with various body parts; and especially a fetus, of the same age, with Chest & head so entangled in, & twinned with the aforesaid substance, that neither Head, nor Chest may be discerned.

A. Two ill-formed & twisted legs, to which is joined

B. A severely bent arm, with a hand.

C. A foot emerging from the upper part of this substance, and having no association with the above-said parts.

D. Roundish bodies, membranous, cartilaginous, and cellular.

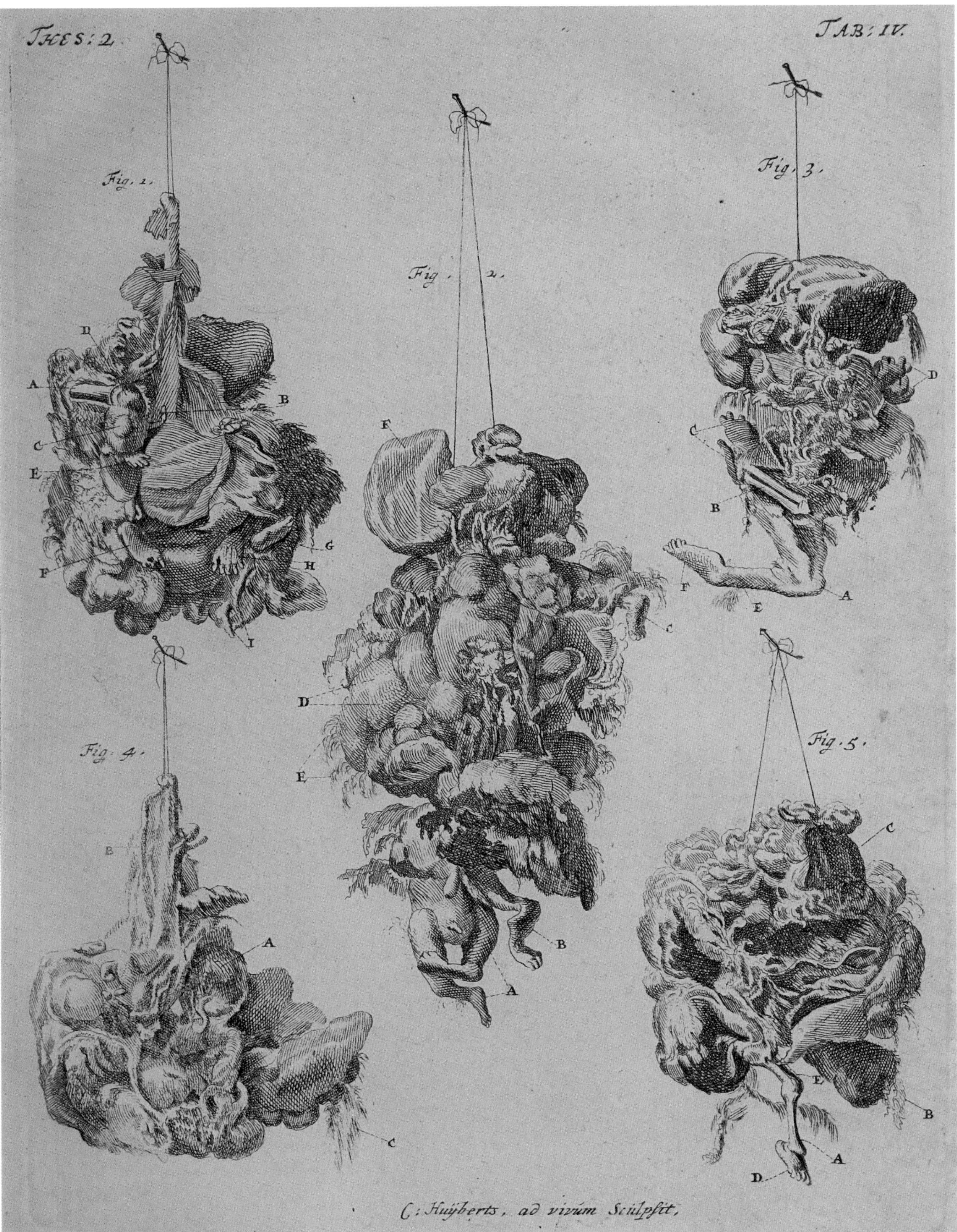
THES: 2.
TAB: IV.
Fig. 1.
Fig. 2.
Fig. 3.
Fig. 4.
Fig. 5.
A
B
C
D
E
F
G
H
I
C: Huijberts, ad vivum Sculpsit,

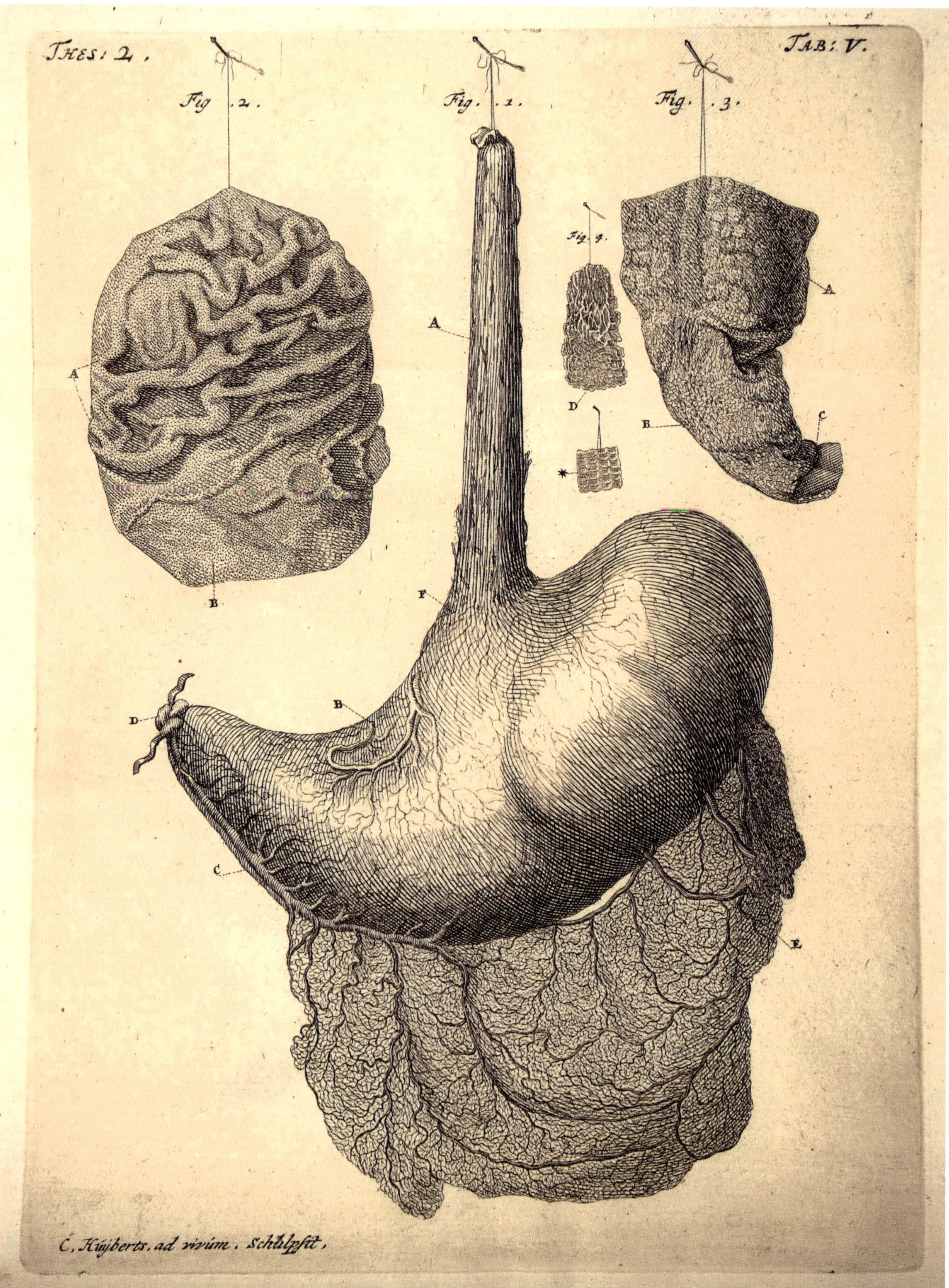
THES: 2.
TAB: V.
Fig. 2.
Fig. 1.
Fig. 3.
Fig. 4.
A
B
C
D
E
F
*
C. Huyberts. ad vivum. schulpsit.

E. Fibers, or blood vessels; these could not be seen, were they not filled with the wax-like substance.

F. A large membrane, full of a slimy substance.

G. Something comparable to a belly, from the lower middle part of which hangs something like a penis.

Fig. 3. Displays a smaller sample of the aforesaid substance, out of which appears:

A. The thigh of a three-month old fetus.

B. The Shin & calf bones of another fetus, denuded of flesh & membrane, and running crosswise.

C. A foot from the same fetus, so it seems, even though it is slightly separated from the said calf & shin bones. Below the said foot appears an unexpected, solitary digit, which I have also marked as C, but with its own line.

D. Two feet without shins or thighs, emerging across from the aforesaid body parts.

E. A shin.

F. A foot.

Fig. 4. Explanation, in which a sample of the same substance is presented, without any visible body parts.

A. Particles, some cellular, some cartilaginous.

B. Membranous particles.

C. Vascular particles.

Fig. 5. Also illustrates a sample of the above-mentioned substance, from the bottom of which a complete leg is extended.

A. A shin bone, stripped of flesh.

B. Vascular particles.

C. Membranous cellular particles.

D. A foot.

E. A thigh bone.

Table V [see image on facing page]

Fig. 1. Shows the stomach of an infant, with the Neck & Caul attached.

A. The Throat, with its long muscular fibers.

B. Arteries running through the upper part of the stomach.

C. Arteries distributed throughout the lower part, whose major branches look as though they were abruptly lopped off.

D. The exit, or pylorus, of the stomach.

E. The caul, provided with myriad blood vessels: but no perforations.

F. The upper mouth, or entry to the stomach.

Fig. 2. Displays a portion of the stomach of person of advanced age, turned inside out.

A. Interior folds projecting quite a bit beyond the interior surface of the stomach.

B. Countless pores.

Fig. 3. Represents a piece of human stomach, turned inside out, severed from above the pylorus.

A. An area where no folds appear, but the inmost surface sprouts forth something like tiny mounds.

B. Interstitial cellular tissue, which is a new observation.[110]

C. The pylorus.

Fig. 4. Shows a portion of stomach in which the said interstitial cellular tissue is represented as seen under a microscope, much larger by far than actual size.

Table VI. [see image on facing page] [111]

Explanation.

Fig. 1. A sample of the portal vein[112] from a calf's liver.

Fig. 2. A sample of veins & arteries from a human womb.

Fig. 3. A small Branch of the vena cava[113] from the liver.

Fig. 4. A branch of human portal Vein.

Fig. 5. A branch of spleen Artery from a calf.

Fig. 6. Shows a piece of the inner lining of the womb of a pregnant sheep, throughout which myriads of newly found wormy vessels are distributed; these are transparent,[114] and prepare juices for nourishing the fetus, while lodging in the womb.

Fig. 7. Shows the worm-like course of the blood vessels that constitute the greatest part of the substance of the kidney.

Note: Nothing is represented here but the course of the said vessels; it is my intention to show quite clearly for posterity the tips of the individual branches, for the individual extremities resemble husks of grain; but in the liver, & also the spleen, they resemble brush bristles.[115]

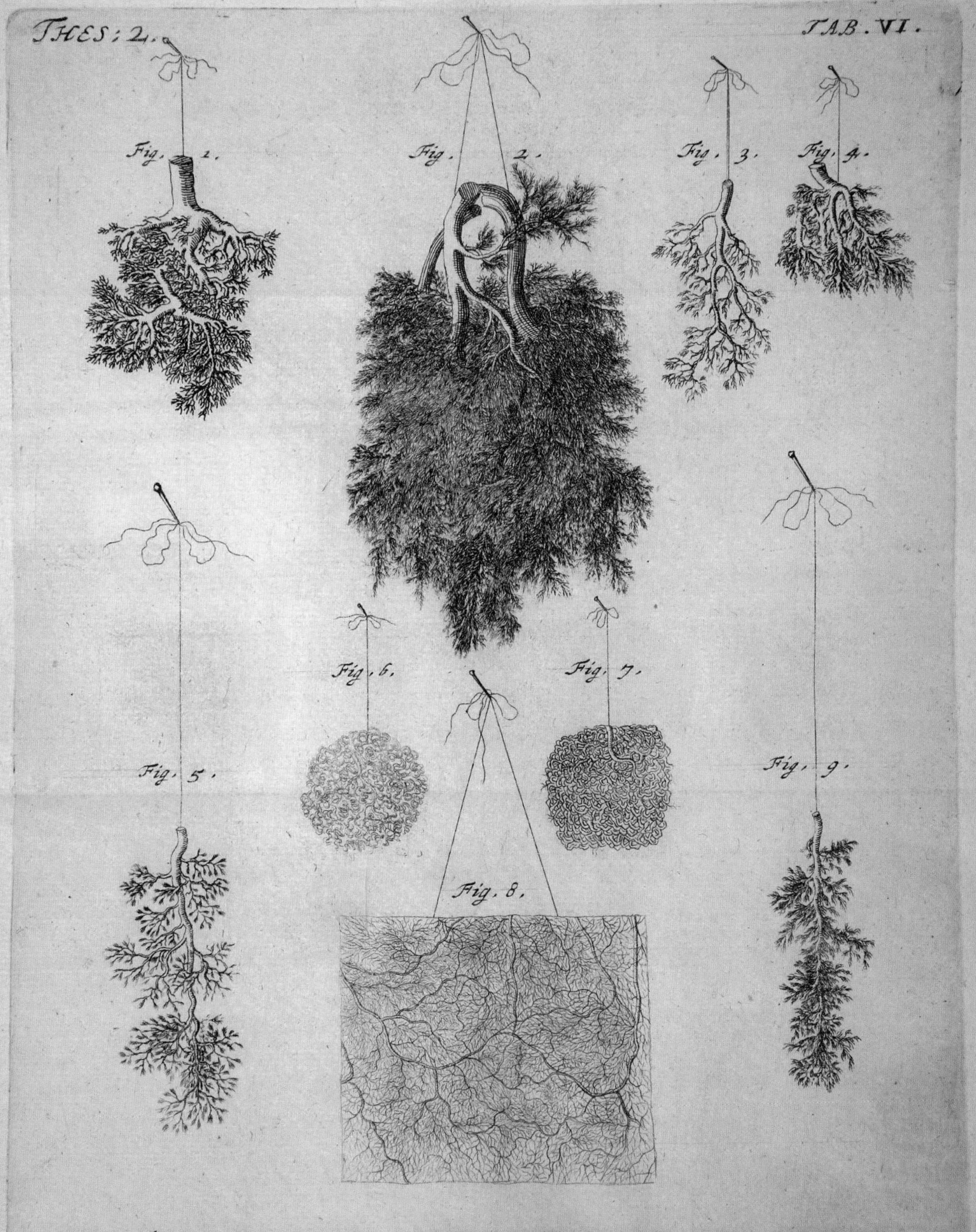
THES: 2.
TAB. VI.
Fig. 1.
Fig. 2.
Fig. 3.
Fig. 4.
Fig. 5.
Fig. 6.
Fig. 7.
Fig. 8.
Fig. 9.
C: Hüyberts,
ad vivum Sculpsit,

Fig. 8. Shows a sample of the lining of the chorion[116] of a calf, whose blood vessels are filled all the way to the furthest tips.

Kind Reader, here you are observing correctly the most minute branches, but in the actual object itself, when observed in poor lighting, these said extremities misleadingly resemble specks.[117]

Moreover, I consider it worthy of consideration that the extremities of these vessels do not at all correspond in shape with the vessels of the liver, spleen, kidney, or uterine placenta.

END.

Frederik Ruysch

Professor of Anatomy & Botany

The Third Anatomical Cabinet

With Engravings

Amsterdam
Johan Wolters
1703

TO THE MOST NOBLE AND ILLUSTRIOUS
JOSEPH PITTON DE TOURNEFORT,
EASILY THE CHIEF
BOTANIST OF HIS GENERATION,
MEMBER OF THE ROYAL ACADEMY OF SCIENCES,
DOCTOR OF THE MEDICAL FACULTY
OF PARIS,
PROFESSOR OF BOTANY
AT THE ROYAL GARDENS,
AND
MY OLD FRIEND,
I DEDICATE THIS THIRD CABINET
FREDERIK RUYSCH

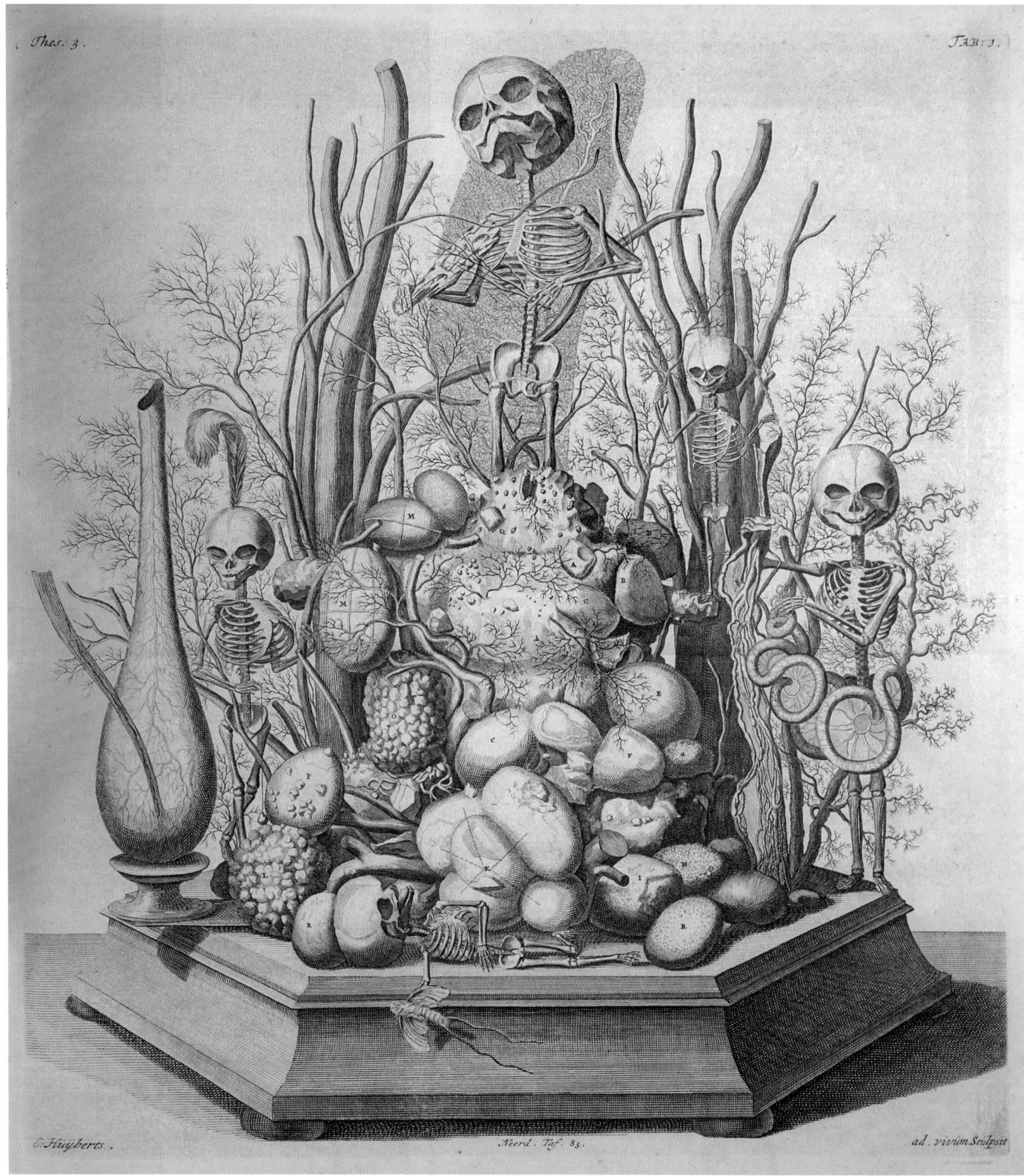

Thes: 3.
TAB: 1.
C. Huijberts.
Neerd. Taf. 83.
ad. vivum Sculpsit

The Third Anatomical Cabinet.

Made from India-Wood in variegated color, it contains the following:
[see image on facing page]

First, in the center bottom, we encounter a Walnut Pedestal, on which rests an artificial cliff-scape, made from human stones; the base of this is natural stone, which supports all manner of human stones, & in such number that the entirety of the natural stone vanishes from sight.

This Scene surpasses in size the one discussed & depicted in the first figure of The First Cabinet. The Stone (see Table I, Cab. 3) is graced with this caption by a Poet of great name, the most Illustrious Peter Francis:[118]

What is mankind? Or rather, what remains
When dirges cease and we are laid to rest?
Behold: Bare bones or nothing will you see.
A mound of kidney stones—a waste of rocks—
That teaches that the life of man is woe.
This truth in figures he depicts, in words
He explicates—Professor Ruysch, the pride
Of Amsterdam, the glory of physicians![119]

At the peak of this cliff is found the skeleton of a fetus about four and a half months premature, whose knees are bent, and head upturned, as though it would behold heaven. In its hand, it clutches a piece of bone, consumed with decay, & expelled from its thigh bone, & thus it seems to express the miseries of mankind.[120] It has this caption:

Ah, fate! Ah, bitter fate!

Beneath the knees of this skeleton one can see a large fragment of human stone, with a remarkable cavity, in which it holds yet another stone, like a dense pit: And this aforesaid fragment is about as thick as a writing pen. (See A.)

Note: The stones that make up this scene reveal that some human stones are composed of thin layers, while others, to the contrary, are of quite thick layers.

For this reason, it is clear that Lithotomists are sometimes wrongfully blamed when stones shatter during an operation, otherwise correctly performed, for the matter is out of their hands; for stones made of thin layers are easily broken during extraction.

Beside the aforesaid Skeleton stands a Skeleton about three months old, bearing in one hand a most delicate Branch of Arteries, filled to the very tips with a red, wax-like substance; while the other holds an artificial stalk of the tenderest grass, made of fibers from a woman's urinary Bladder. Nature has covered these fibers on all sides with so many grains of fine human stones that hardly anything but these sandy Stones can be seen.

Note: Because the urinary Bladder that provided the said fibers for the stalk of grass was composed quite entirely of this sort of stony fiber, the whole Bladder looked to me at first sight as though it were made of stone—or, rather, of grains of sand.

I placed in Embalming fluid a sample of this Bladder large enough to preserve its natural composition.[121]

The aforesaid little skeleton appears to be indicating the composition of its head with its stalk of grass, for it touches this to its forehead, as if to expose the error of Anatomists who maintain that the forehead, & other bones of the head, at this age, are cartilaginous, when, to the contrary, they are membranous.

Furthermore, one can clearly see in this small Skeleton the locations where the bony matter first appears in the head; from here, more bone will follow.

In front of this aforesaid Skeleton are three large human stones, differing from each other in shape, color & substance. (See Letters B. C. D.)

Beneath these, a large human stone comes into view (marked as Letter E) so closely resembling an Oriental Bezoar stone[122] in its smooth & shiny surface and in its color & shape that it would be difficult to distinguish them by sight.

Beneath that, one may see two stones that could not be more different (marked as F & G); one of the pair extracted by Lithotomy has a grainy texture; the other has a smooth & almost polished surface, in which the extracting instrument has left a mark.

The above-said Pedestal with the cliff-scape is also adorned, behind the upper skeleton, with a neatly prepared calf Spleen, whose Artery has been filled with a red waxy material, & Furrows can clearly be seen in the Veins. How great a difference there is between a human & calf Spleen, I have illustrated in *Anatomical Letters, Problem* 4.[123]

Near the base of the Rock display, one may see a human stone, large & gray (marked with the letter H.); these are the worst sort of stone, for I have observed, that among those gray stones, most (if not all) have a brittle exterior shell; gradually fracturing one after another, these outer scales are often passed along with urine, while the core remains: when they are removed by Lithotomy, the patient seldom avoids the Grave.

I recall that I once observed this in a certain Youth, who was passing black fragments, that resembled the husks of black beans, and at first they were soft and passed with the urine without great inconvenience or pain while making water, but soon they acquired a stony & brittle nature.

I.[124] A large stone composed of thin layers.

At the base of the Cliff is found a stone (marked with the Letter K), of prodigious size & marvelous shape, cut from a bladder, *post mortem*, by my son, and compounded from six round stones, possessed of black shells; each of the constituent stones is larger than a Nutmeg. Above this composition is also found a seventh stone, which has not joined to the others, but has been placed on top of them, and is of the same size.

L. A very large, Blackberry-shaped stone.

Next to this Skeleton is found also a third Skeleton of a four-month-old human fetus, holding the small gut of a tiny Sheep fetus, which has been filled with a blackish wax-like substance, representing the excrement particular to a fetus,[125] and which is held in shape by its mesentery, itself furnished with wax-filled arterioles: the other hand clutches the Vine-like Vessels from a pair of human Testicles, filled with a red, wax-like material. It bears this caption:

The hour that first gave life already has plucked it.[126]

For consideration: of all the Skeletons embellishing this & the other Rock displays, none is supported with a brace of either iron, wood, or copper, as Anatomists usually do, but each stands upon its own feet.

M.M. Two combined stones, above which is found a stone of moderate size (marked as N).

O. A large stone, most greatly resembling a Blackberry.

P. —with a rough surface.

At the foot of this cliff lies a Skeleton, about three and a half months old, whose hand holds an *Hemerobius hoefnagelii*, or mayfly, which is called *Ephemeris* by Dortmann, and which is said to be newly born in the morning, youthful at noon, and worn out with age & dead by evening. And just so, the brevity of our life is expressed by this insect, held in the hand of the said Skeleton. It has this caption:

Like the summer grass, my time was brief,
Rapidly I rose, and rapid fell.[127]

Note: the gestures of this little Skeleton are most elegant, since attention to nature has not been neglected.

Q.Q. Two of those angular stones of which I made mention in Cabinet 1, Table 1, Letter D.D.

R.R. Two stones with quite rough surfaces, one of which is set at the feet of the said Skeleton, while the other is placed at the top of its head.

All the stones were either excreted from the urinary path or cut out, *post mortem*, from the Bladder or kidneys.

Furthermore, one may see here the Vaginal Tunic[128] of the human Testicle, supported by its own pedestal, that rests upon the said Rock pedestal, and the said Tunic is taut, as though inflated, and is well endowed with Cremaster[129] muscle and blood vessels.

Finally, across from the said Vaginal tunic, one comes upon a fifth human Skeleton of about three and a half months, holding a little stone, coughed out of a human lung & and hanging from a stand of hair. It has this caption:

Alas, harsh life bears down upon us wretches![130]

Among all these stones are interspersed arterial Branches, representing tree branches stripped of leaves; in place of blood, they have been artfully filled with a deep-red waxy substance, & they help hold together the covering of little stones and provide greater stability to the overall Cliff scene. The fine delicacy of the tips of the said branches is clear from the accompanying figure.

[...]

No. VI. [see image on facing page] The fetus of an Ethiopian[131] girl of about six months, right hand holding a branch of Wind-pipe, of vascular, or rather pulmonary substance, removed with such dexterity that the tips show quite clearly; to be sure, there is quite a difference between this object & the figures others have published elsewhere! (See Table 2, of this Cabinet.)

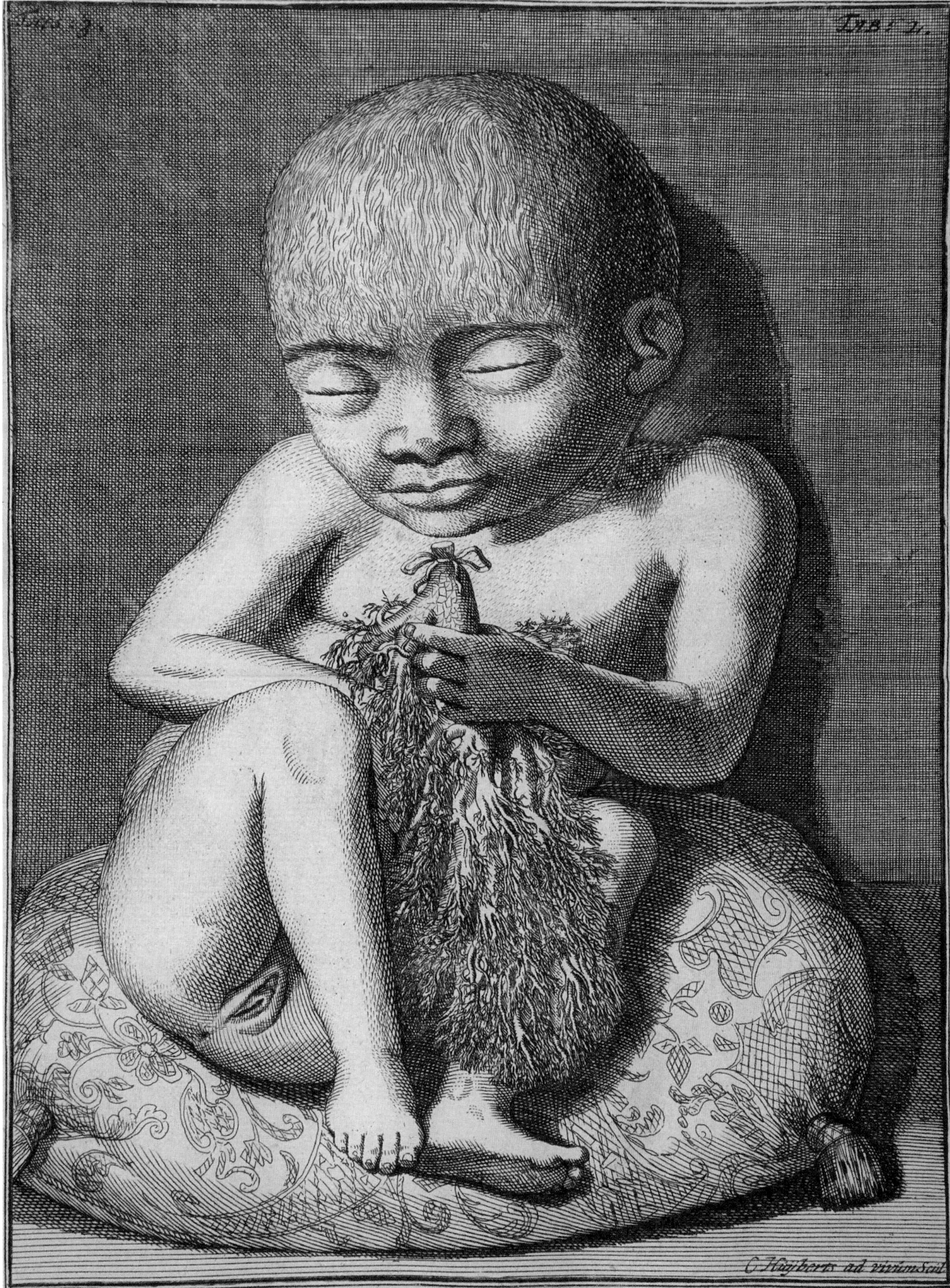
TAB. 2.
C Huijberts ad vivum Sculp.

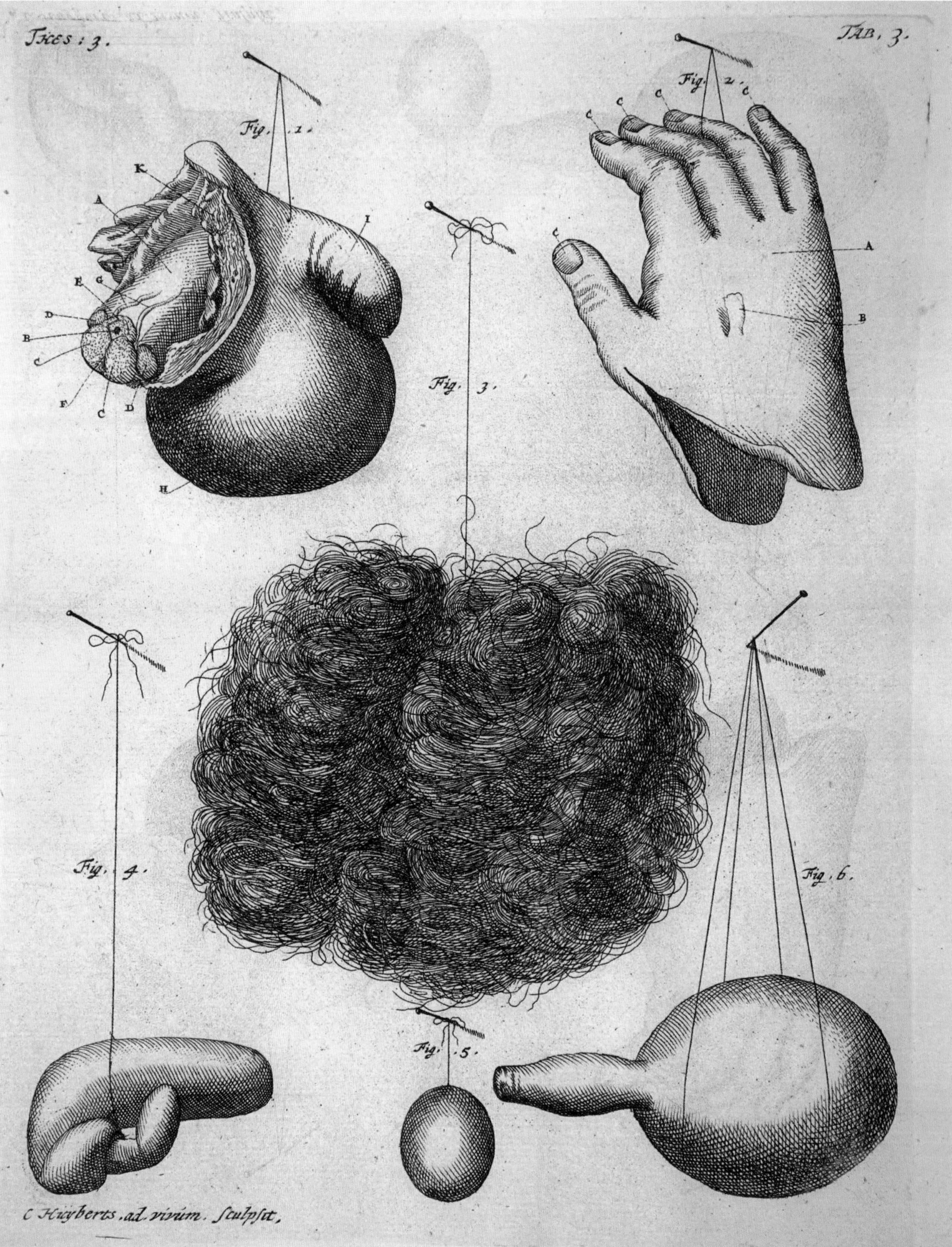
Thes: 3.
TAB. 3.
Fig. 1.
K
A
I
E
G
D
B
C
F
C
D
H
Fig. 2.
C
C
C
C
C
C
A
B
Fig. 3.
Fig. 4.
Fig. 6.
Fig. 5.
C Huyberts. ad. vivum. sculpsit.

ANATOMICAL CABINET III: EXPLANATION OF THE FIGURES

EXPLANATION OF TABLE THREE [see image on facing page][132]

Fig. 1. Represents a Child's Penis with Scrotum.

A. The interior part of the Penis

B. The Urinary Passageway along the back of the penis, running contrary to the law of nature.

C.C. The two larger nervous bodies, or the greater nerveo-spongeous bodies.[133]

D.D. Muscles for constricting the Penis, severed at the origin.

E. The corpus spongiosum,[134] or the lesser nerveo-spongeous body, circling the urethra, contrary to nature.

F. The septum of the Penis.

G. Parts of the Arteries running along the back.

H. The Scrotum.

I. Part of the Penis hanging outside the body.

K. Fat, with which the upper part of the scrotum abounds.

Fig. 2. Depicts the Epidermis of a child's hand, separated from the *corpus reticularis*[135] & the cuticle so it comes to us intact, like a white Glove.

A. The back of the epidermis.

B. A small fold of Epidermis, separated here from the back.

C.C.C.C.C. Fingernails, which, during the separation from the skin, did not come off the Epidermis, and are in fact still attached to the Epidermis.

Fig. 3. Displays a bunch or clump of hair discovered in an Atheroma[136] inside a human bowel.

Fig. 4. Shows a chicken egg, with an ordinary hard & white shell, that perfectly resembles the blind gut with appendix.

Fig. 5. Presents a tiny Chicken Egg found inside another chicken egg.

Fig. 6. Illustrates a chicken Egg that resembles a Child's Penis with scrotum, & not even lacking the foreskin.

EXPLANATION OF TABLE FOUR [see image on facing page]

Fig. 1. Shows an infant's face, with the forehead removed.

A. A bundle of Arteries obscuring & covering over the wound I made in removing the forehead, so that I might lessen any disgust; these are arteries that spread throughout the brain, & here they have been filled with a red, waxy material.

B. The Lips, with their covering, or Epithelide, removed, so that the Papillary nerves that constitute the Lips should stand out & come into view.

Fig. 2. Shows a human Kidney, seen from the outside.

A. The top layer of the Kidney, throughout which spread countless worm-like blood vessels.

B. The Renal Artery, severed.

C. The renal vein, severed.

D. A piece of a pelvic ureter pushing out beyond the body of the kidney, through which many arterioles are disseminated.

Fig. 3. Shows a bisected human Kidney, so that the inside lies open.

A. Indicates that the tips of the many arterioles devolve into the Ducts of Bellini, & are continuous with them.

[...][137]

B. A small piece of the surface layer, from the outside.

C.C. Renal papillae, perforated with many holes, or mouths, for transporting urine.

D. The Ureter.

E.E. Openings of the renal Pelvis.

Fig. 4. Traces the fibers of slender nerves, subdivided into countless tinier fibers, as slender and delicate as a spider's web.

Fig. 5. A sample of Nose, cut lengthwise, taken from an infant.

A. Part of the Forehead bone.

B. A cartilaginous Rooster's Crest,[138] of tender age.

C. Part of the Nose.

D.D. Bristles inserted into the nasal canals.

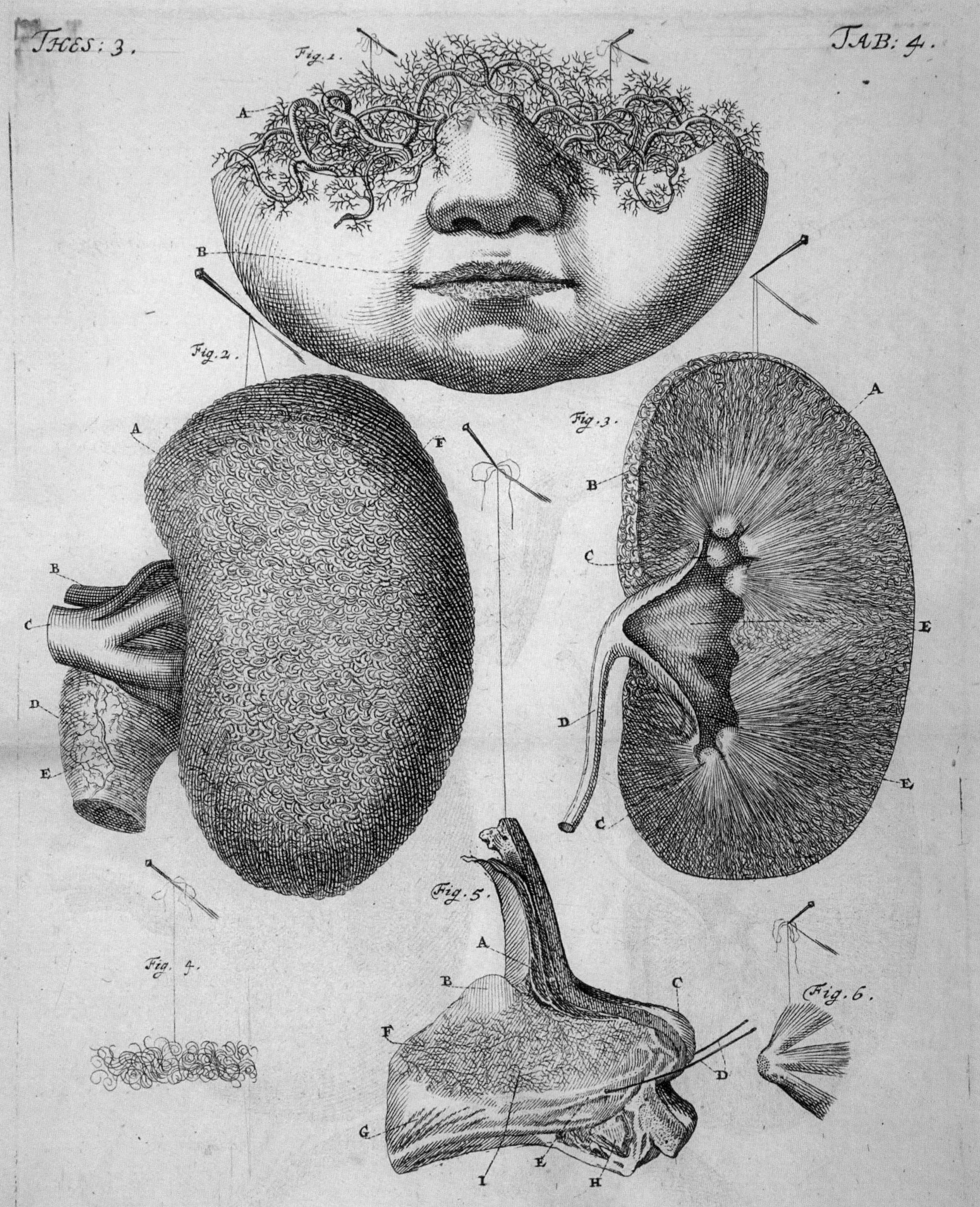
THES: 3.
TAB: 4.
Fig. 1.
A
B
Fig. 2.
A
F
B
C
D
E
Fig. 3.
A
B
C
E
D
E
C
Fig. 5.
A
B
C
F
D
G
E
H
I
Fig. 4.
Fig. 6.
C. Huijberts.
ad vivum sculp.

E. One of the nasal Canals.

F. Blood vessels distributed across the thick covering of the Septum; these are primarily Arterioles.

G. Oblique furrows.

H. One of the Incisors.

I. Connective tissue[139] immediately surrounding the cartilaginous septum of the Nose.

Fig. 6. Shows a Renal papilla, composited from three different samples.

END.

Frederik Ruysch

Professor of Anatomy & Botany

The Fourth Anatomical Cabinet

With Engravings

Amsterdam
Johan Wolters
1704

Explanation of the Tables

The First Table [see image on following page]

Fig. 1. Displays half of a human Kidney, dissected so that the crawl of the vessels, particularly the blood vessels, can be seen more clearly that it was in the preceding Cabinet (Table IV, Fig. 3), where I was more intent on showing the conjoining of the arterioles with the ducts of Bellini; however, in this figure, I wish to express very distinctly the worm-like course of the blood vessels through this part of the Kidney.

A. The exterior face of the Kidney, through which the blood vessels observe their worm-like crawl, just as I expressed in the previous Cabinet but more clearly.

B. The interior face of the Kidney, where the blood vessels observe a worm-like course no less than in the exterior face.

Note 1: The said crawl of the vessels is expressed so finely, in this figure, that it should be easily visible to the naked eye.

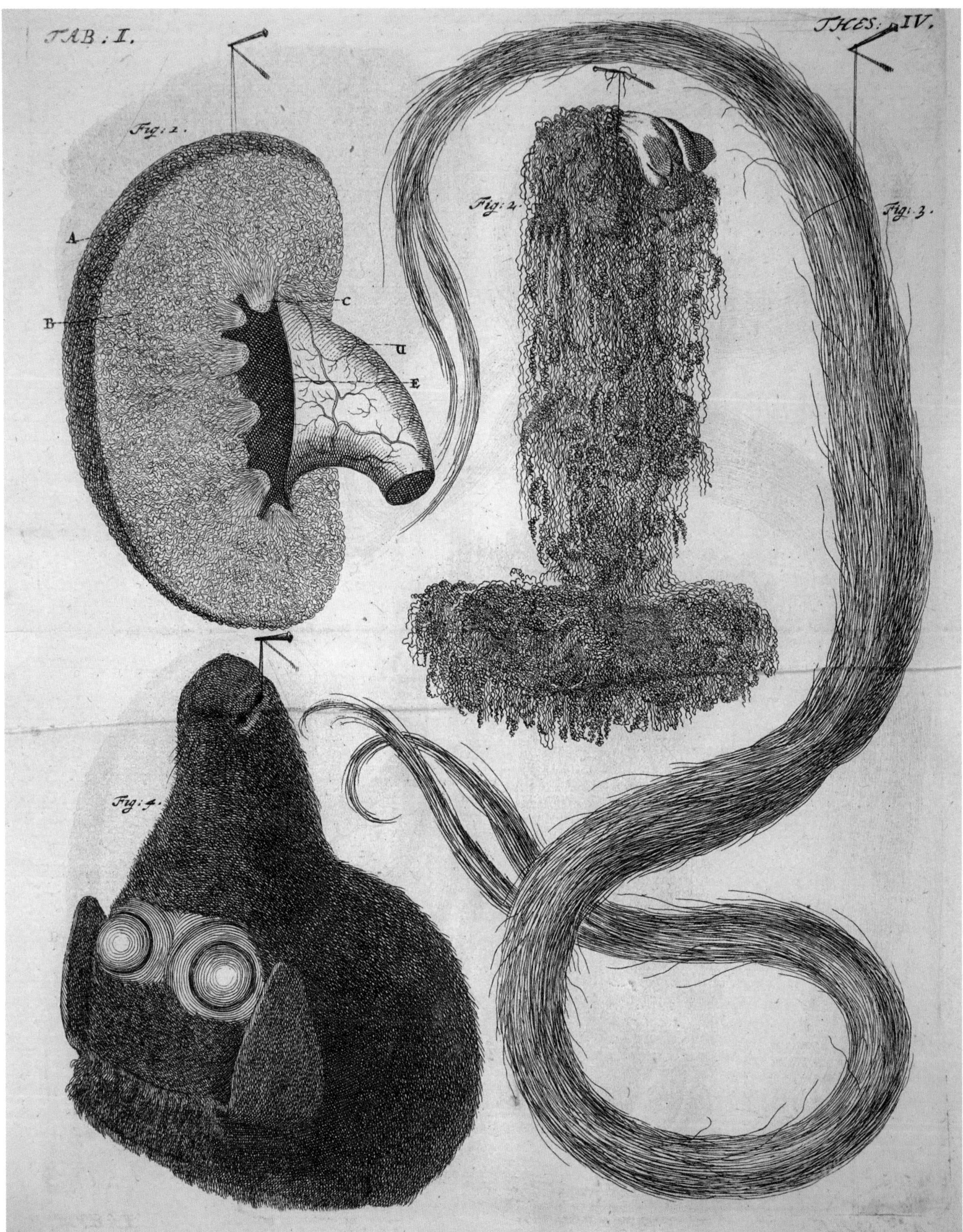

TAB: I.
THES: IV.
Fig: 1.
A
B
C
U
E
Fig: 2.
Fig: 3.
Fig: 4.

Note 2: The holes, which I mentioned in the preceding Cabinet, do not appear here, since they disappear when the Kidney has been dried.

C. A Renal Papilla.

D. The Renal Pelvis.

E. The hollow of the pelvis, into which the papillae express urine in drips.

Fig. 2. An unloosened human Testicle, which is nothing but bunches of the most delicate vessels.

Fig. 3. A genuine Polish Plait.[140]

Fig. 4. The head of a sheep fetus, deformed & mouthless, or at least without a formed mouth.

Table II [see image on following page]

Shows a bovine urinary bladder, turned inside-out, bursting with tumors.

Fig. 1.

A.A.A. Said tumors, petrified.

B. Soft tumors.

Fig. 2. Shows IV stones that fell out of cavity of the said bladder.

Fig. 3. Indicates a few stones that fell out of the inner coating of the bladder, some on their own and some loosened by my fingers.

The Third Table [see image on page 163]

Displays a human Heart, completely filled with a red, waxy substance, so the true shape of the Heart & the Auricles can be seen.

Fig. 1.

A. The Heart, seen from the flat face, which lies against the diaphragm.

B. The Trunk of the Coronary vein with its branches.

C.C.C.C. Branches of the Coronary Artery, whose extremities are like fine down feathers, even though they are filled with a waxy substance.

D. Part of the right Auricle (only because, in this position, the entire auricle does not meet the eye), through which countless blood vessels are dispersed.

E. Part of the remaining diaphragm.

F. The descending trunk of the vena cava, passing into the diaphragm.

G. The membranous sac, which is, as it were, the foundation of the Pulmonary Vein, &, with the left auricle, constitutes a communal cavity.

H. The trunk of the great Artery, from whose back emerge three branches, rising upward.

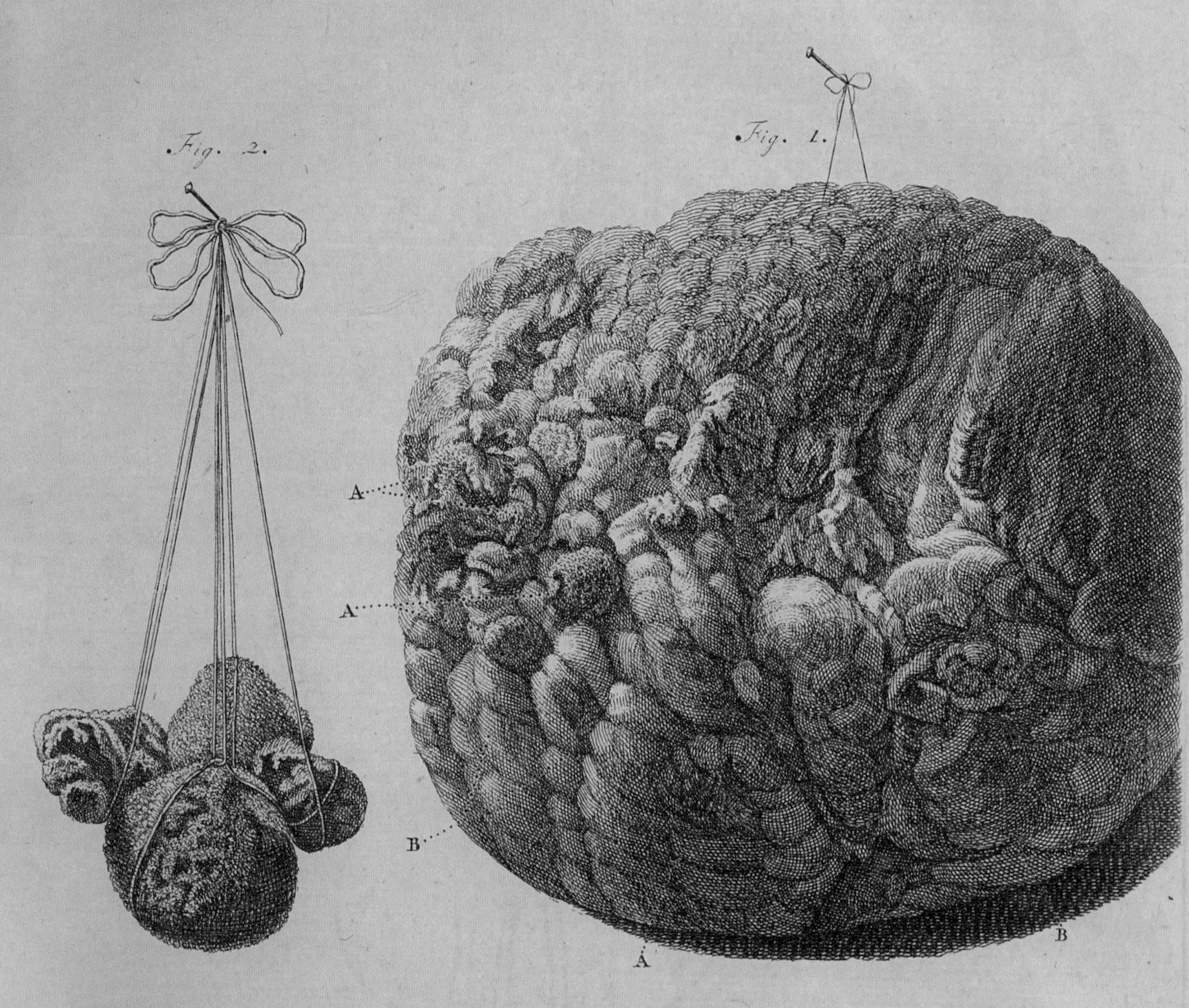

Fig. 3.

A. de Blois ad vivum fecit.

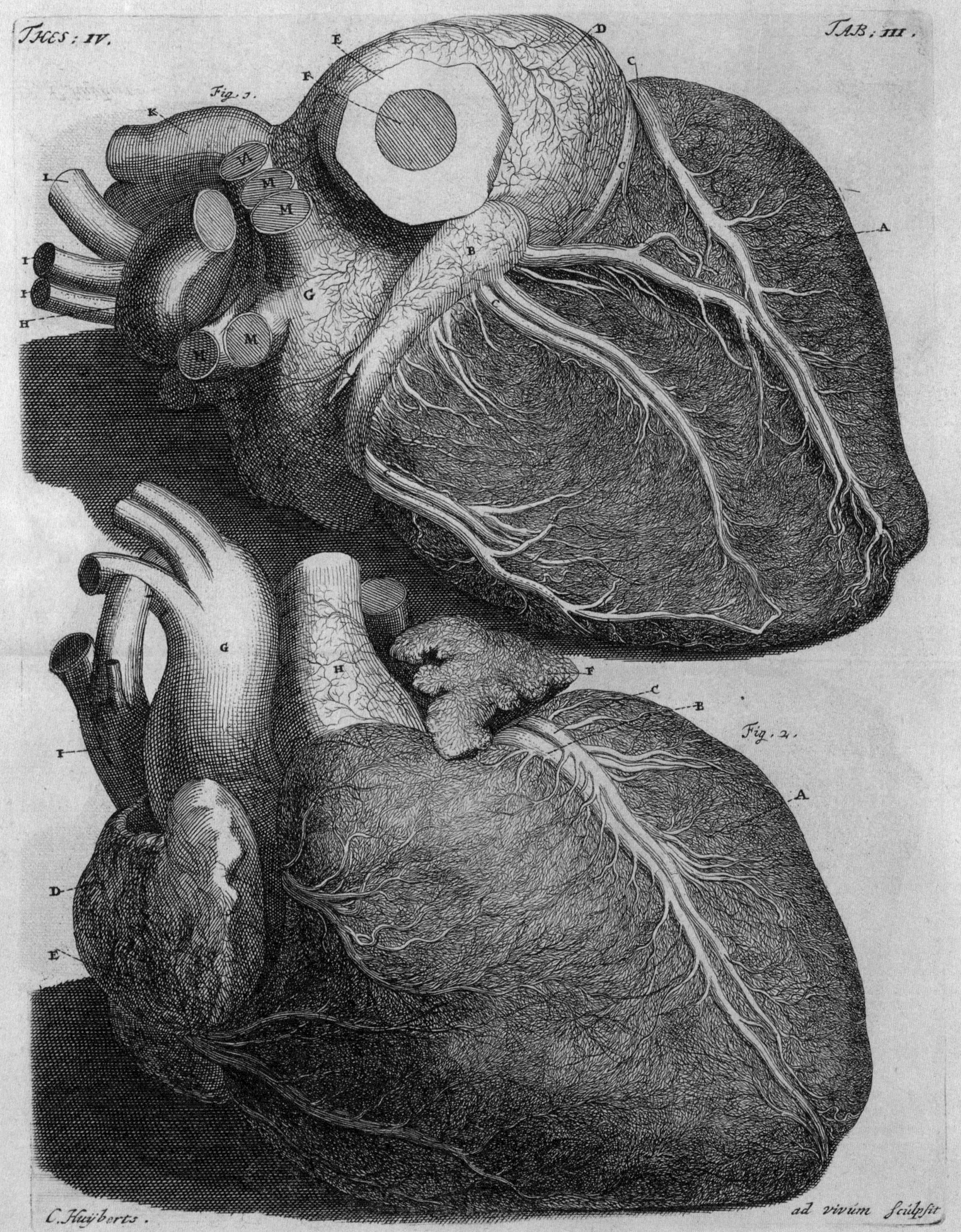
THES: IV.
TAB: III.
Fig. 1.
A
B
C
D
E
F
G
H
I
K
L
M
Fig. 2.
A
B
C
D
E
F
G
H
I
C. Huijberts.
ad vivum sculpsit

Note: The said branches appear to us just this way (with the Heart so positioned) when looking at the flat surface.

I.I.I. Three branches of the great Artery rising upward.

K. The rising trunk of the vena Cava.

L. Part of the left Auricle.

M.M. Branches of the pulmonary Vein.

N. One of the branches of the pulmonary Artery: other branches do not come into view here, with the heart placed as it is.

Fig. 2.

A. The same human Heart, seen from the rounded face, & in its natural location.

B. Branches of the Coronary Vein.

C. Branches of the Coronary Artery.

D. Trunk of the pulmonary artery.

E. Trunk of the ascending Vena Cava.

END.

Frederik Ruysch
Professor of Anatomy & Botany

The Fifth Anatomical Cabinet
With Engravings

Amsterdam
Johan Wolters
1705

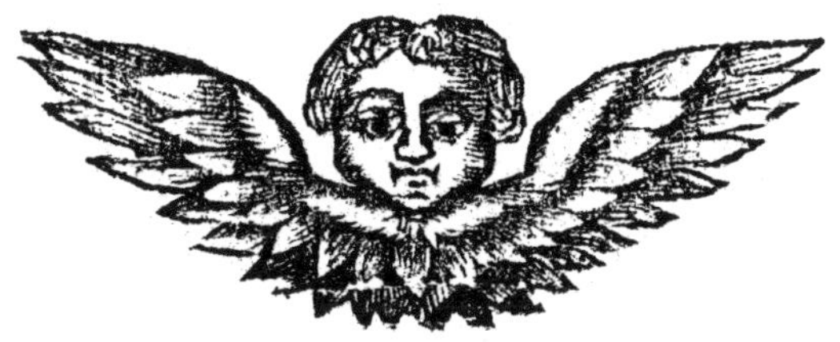

EXPLANATION OF THE FIGURES OF THE FIFTH CABINET.

Table I [see image on following page]

Figure 1. Presents an illustration of human afterbirth.

A. The surface of the uterine Placenta, which faces the womb & is enveloped in a shaggy covering.

B. A shaggy portion of Chorion,[141] hanging from the placenta.

C. A thin portion of Chorion, separated by a thicker & a thinner portion, & which I take for the Allantois[142] (though it is less craggy).

D. A piece of loosened placenta, clearly revealing that its composition is e ntirely of vessels, with no trace of the much-vaunted glands.

E. The umbilical cord.

F. A hairier piece of chorion, spreading over that part of the Placenta that faces the Womb, & is connected to it.

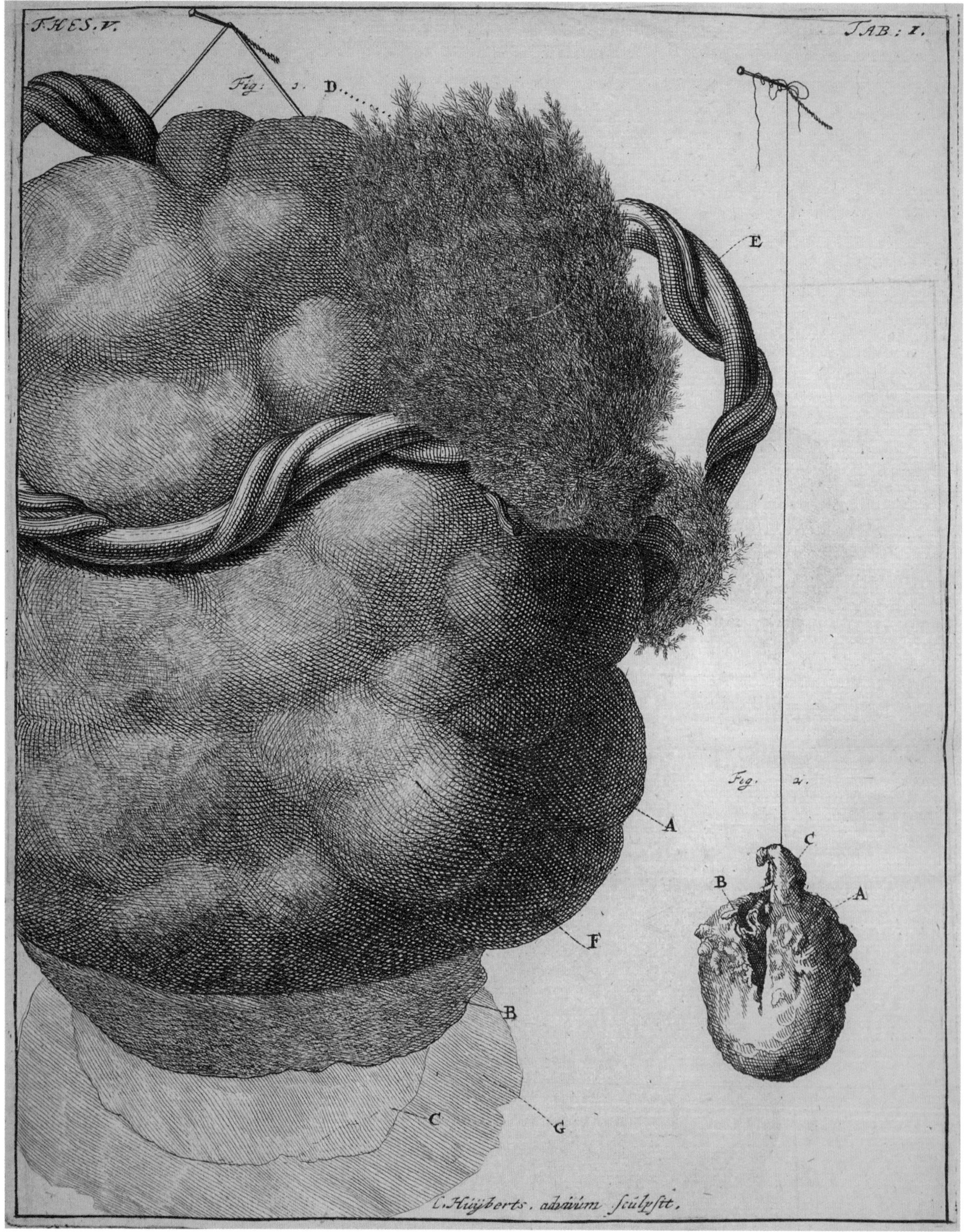
THES. V.
TAB: I.
Fig: 3.
D
E
Fig. 4.
A
C
B
A
F
B
C
G
C. Huijberts. ad vivum sculpsit.

G. A piece of the Amnionic coat.

Fig. 2. Displays an Image of the uterine afterbirth of an Embryo, about two months old, that following the destruction, or liquefaction, of the Embryo, remained in the womb for some space of time, whence, being compressed more and more, over time, by the womb, it hardened, & thus represents nothing less than afterbirth.[143]

A. The outer surface of the Placenta.

B. A hole made by my forceps, in order to bring the hollow into view.

C. A piece of the coagulated blood, which stuck to the neck of the womb, & resembles a tiny foot.

Note: When an Embryo is reduced to nothing, or dissolved, the placenta & membranes are often covered with much blood, whence the afterbirth seems to be larger in proportion to the Embryo.

Table II [see image on following page]

Fig. 1. Shows the Head of a thigh bone sawn in two.

A. The smooth, exterior area, encrusted with cartilage.

B. The interior osseo-spongeous area.

C. A pocked exterior area, covered with no cartilage.

D. A bony layer, quite thin, & hardly visible in one area.

Fig. 2. Shows the upper part of the thigh bone of a young boy, missing the two *apophyses* (a term I consider spurious), or trochanters.[144]

A. The said upper part, cut with a saw.

B. A small projecting part, from which the lesser trochanter became detached, after the bone was cleaned by boiling.

C. The neck of the thigh bone, whose head was sawn off, to best reveal its interior osseo-spongeous substance, & the thinness of this layer.

D. The bony layer, which on one side is extremely thin & so provides an opportunity for fractures in this place.

E. The osseo-spongeous substance of the thigh.

Fig. 3. Displays a lesser Trochanter, which, after cleaning, was reduced to the actual bone.

Fig. 4. Represents the upper part of the Skull of an infant, in whose hollow the dura mater still remains, with its Arteries, filled with a red wax-like substance.

A. Part of the exterior surface of the Skull.

B. The course of the Arteries through the dura mater.

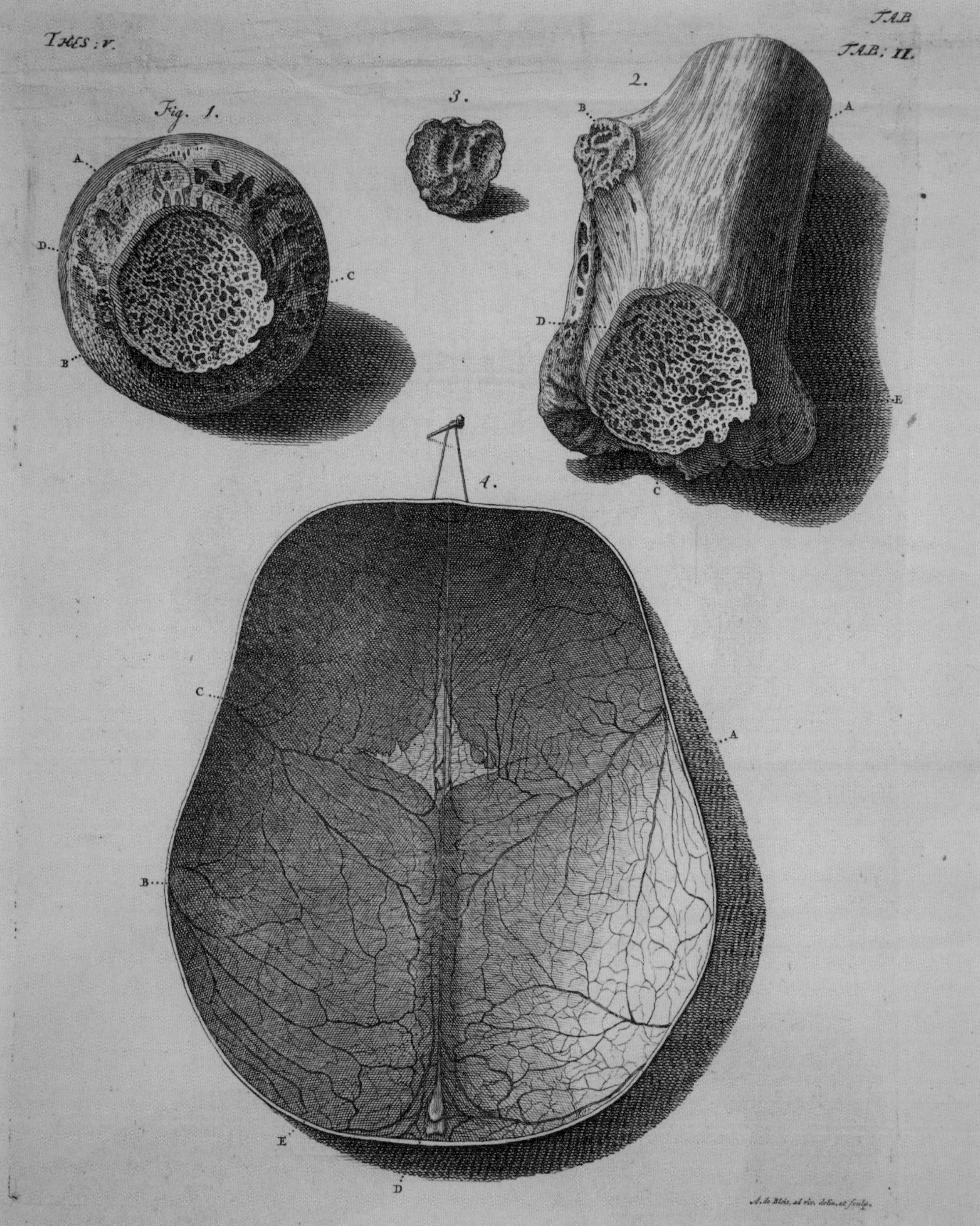
TAB
THES: V.
TAB: II.
Fig. 1.
2.
3.
4.
A
B
C
D
E
A. de Blois, ad viv. delin. et sculp.

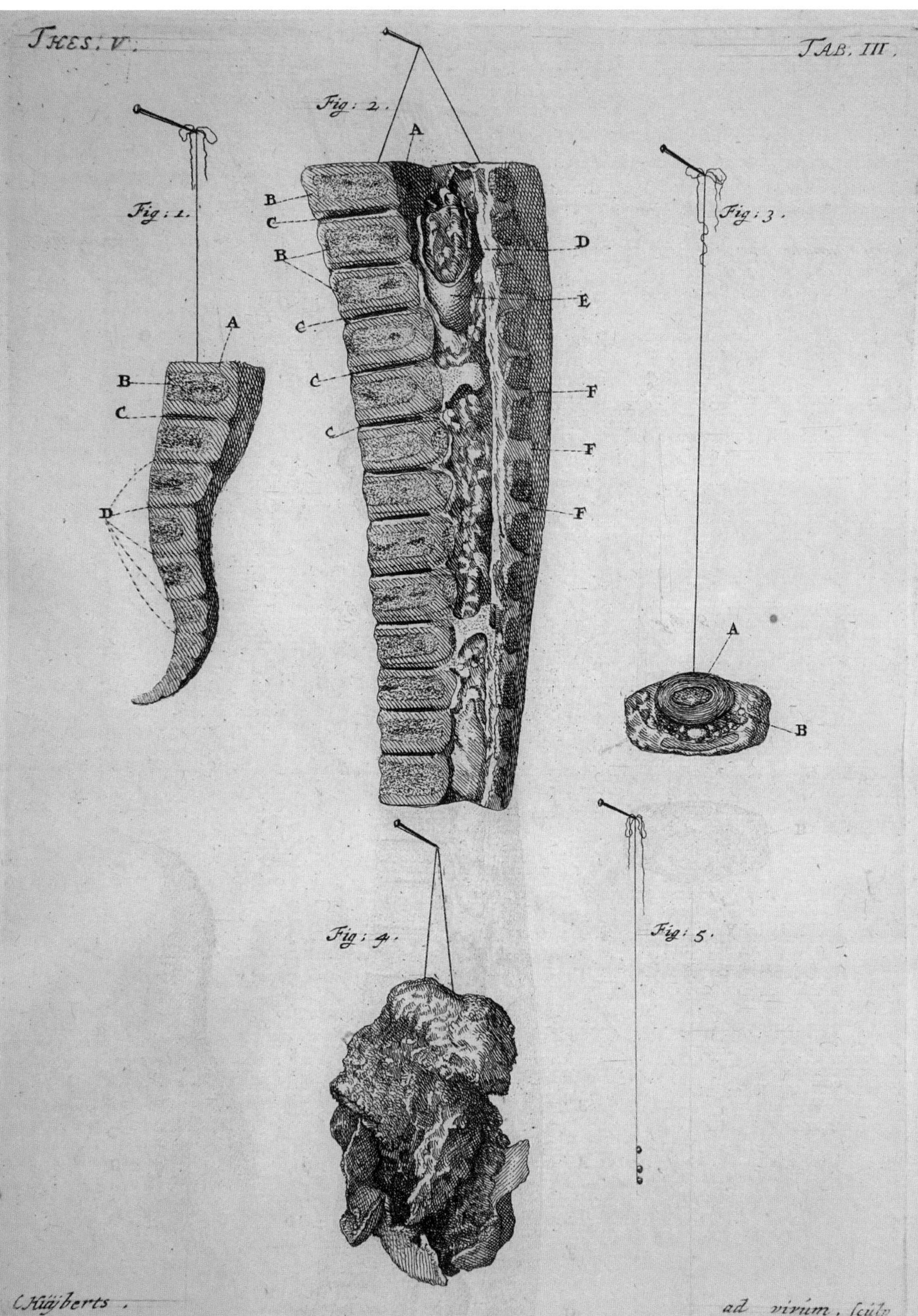
THES. V.
TAB. III.
Fig: 1.
Fig: 2.
Fig: 3.
Fig: 4.
Fig: 5.
A
B
C
D
E
F
C. Huijberts.
ad virúm. sculp

C. Their course along the side of the *processus falciformis*,[145] below the Fontanella.[146]

D. The fold of the *processus falciformis.*

E. The *Falx*.[147]

Table III [see image on previous page]

Shows IV Figures, of which

Fig. 1. Shows a portion of the Sacrum Bone, the Tail Bone & the lowest vertebrae of an infant, cut in half & seen here from the side.

A. A piece of the body of a vertebra of the lower Lumbar.

B. The interior osseo-spongeous part.

C. A hollow between the body of the last vertebra of the Lumbar & the first sacrum Bone.

D. The spaces between the bodies of the vertebrae of the Sacrum bones.

Fig. 2.

Letter A. shows a piece of the Spine, split in two lengthwise, & seen from the side; taken from an infant.

Letters B.B. The osseo-spongeous interior substance of the body of the vertebrae.

C.C.C.C. Hollows between the bodies of the vertebrae.

D. A portion of the Spinal Cord.

E. The covering of the same, continuous with the dura mater.

F.F.F. Cartilaginous spinous processes.[148]

Fig. 3.

Letter A. Outlines the body of a vertebra from the same spine, seen on its own.

B. Ligaments situated between twin bodies of vertebrae.

Fig. 4. Represents a degenerated piece of the Tarsus of bones,[149] softer in substance than bone, on account of *spina ventosa*,[150] whence the entire Tarsus appears as a single body, without any division of the bones.

Fig. 5. Presents three tiny bones, or Stones, found in a human Pineal gland. These are hanging from a hair.

END.

Frederik Ruysch
Professor of Anatomy & Botany

The Sixth Anatomical Cabinet

With Engravings

Amsterdam
Johan Wolters
1705

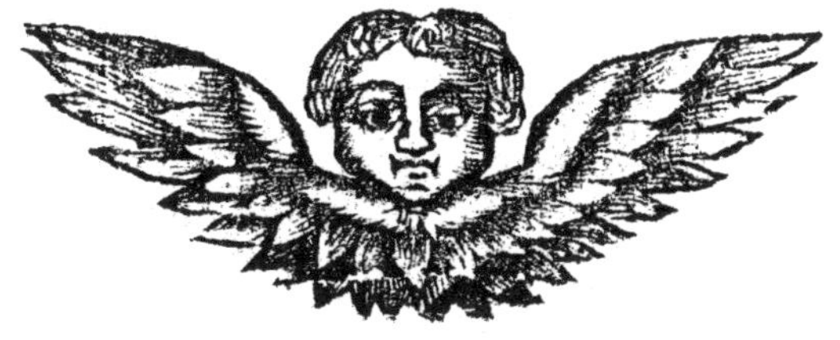

TO THE KIND READER

Having brought to its conclusion the Sixth Cabinet, in which many things entirely new & noteworthy are to be found (as is the case in the volumes that have preceded, & that will follow), I am now of a mind to establish at my home a private Society concerning these matters, throughout the whole course of the year, without interruption, from twelve o'clock until one, or half of two. In this Society I will display every part of a complete body, even to the smallest particles, most of which parts are held in Jars, filled with the clearest liquid, in which they will appear more clearly & with far more detail than they would otherwise. But since many other parts can be seen more clearly after drying, I will also present all those, which I keep in cases, and I will hold a brief discourse about all those preparations. Now, since most of these have been removed from their natural sites, I will demonstrate the location of these parts in a single cadaver; & since some few persons have requested a hands-on demonstration of dissection, for their benefit I shall dissect body parts on two occasions. Once this is done, I will return to the forge, for the benefit of those who have time to stay on, with demonstrations of four-legged animals, Fish, Shellfish, Butterflies, Birds, Sea Creatures, &c., all of which—especially those brought to us from the Indies—are in such great number that I must maintain some fifteen hundred jars, not including the dried specimens I store in cases: since this will require a great deal of time, anyone who wishes to observe with particular attention may do so for an hour each day: everything in the XVI Cabinets that touches on the human body I will try to work through within IV or V weeks.

Frederik Ruysch
Professor of Anatomy

Foreword to the Sixth Anatomical Cabinet,[151]

In which it is demonstrated that it is by no means the Illustrious Master Vieussens,[152] but I who first made the discoveries in the Human body, that the Illustrious Vieussens credits to himself in his treatise, published just recently, under the name *A New System of Blood Vessels in the Human Body.*[153]

It has been some fourteen years since I devised[154] a method of preparing, dissecting, & displaying that excelled common practice, & by means of which the parts of the human body, down to even the most minute particles, may be brought into view, & without which I believe that all those discoveries would have been impossible to demonstrate.

In the year 1696, I undertook the publication of certain Letters on Anatomical Questions, sent to me by various Learned Gentlemen, along with my Illustrated responses.

In those letters, many new discoveries are to be found, since I had by that time devised a method by which human bodies after death might appear to the eye just as though summoned back to life; such is the testimony of all who day after day come to my house to view them. Among these visitors was the Reverend Master G. Papin, of the Society of Jesus, having returned from the East Indies, who graced the cadavers I had prepared with the following verse:

The dead, by your art, Ruysch, now live and teach,
And infants, without language, yet have speech
To tell of secrets long to us denied.
And death itself now dreads that it might die.[155]

Moreover, it was not the practice at that time for Professors of Anatomy, engaging in Anatomical demonstrations, to bring with them preparations suspended in liquid, & to display them at the appropriate moment, until I began to do so; for I considered Anatomical demonstrations conducted on recently deceased bodies to be less useful, if not combined with such preparations, suspended in liquid; in this manner, I made public dissections, a number of times, on the bodies of youths deceased for many years, which is clear from published programs, such as the one that follows:

PROGRAM

Dear Reader

Inasmuch as I have been accustomed each year to perform one or two public Anatomical demonstrations; and it has now been an entire year, since I have been able to obtain a single fresh human subject; I therefore have decided to hold a public Anatomical demonstration at the Anatomical Theater on the first Tuesday of August, 1703; and this will be on the cadavers of those three Boys, who traded their lives for death almost ten years ago, in July, 1695, & whom I displayed to the public in October, 1696.

Through my art, though, these cadavers regained a natural and lifelike appearance, &, what is more, they have become even more handsome over the succeeding years, which will be clear at the appointed time.

The first of them appears as a Boy asleep, with comely face; ruddy cheeks & lips; smooth skin, free of wrinkles; & with no cosmetics or dye—just as was the case in 1696, when, in response to the request of a certain Most Noble Gentleman, I, in the presence of all the spectators, who were gracing those demonstrations with their presence, scrubbed its face with salt, sandy water, soap, & a linen cloth, 'til I had nearly flayed it.

The Brain, Stomach, Guts, Liver, Spleen, Kidneys, Heart, Lungs, and other Viscera, have been so nicely preserved right up until now, that they look better than those parts in many of the living.

I have a mind to present this cadaver dressed in clothing, as has been my custom, without any incisions or lacerations, & I have decided to preserve it thus, untouched.

The second Cadaver, on the other hand, has been dissected by me in public over the aforementioned years, and its Muscles are separated.

The third Cadaver I have dissected in such a way that I may reveal each part individually, & display them to those who are not standing nearby.

Concerning these said cadavers, it should be understood, that everything in them will be displayed—and therefore seen—with much more clarity & detail than is possible in a recently deceased body; it is a well-known truth, that everything in the outward aspect of a living human body that brings joy to the eye is stolen by death, & its essence passes away. Similarly, it is no less true, that the internal parts, after death, retreat considerably from their natural state, and, in fact, the smallest particles flee entirely from our gaze.[156]

However, by this art of ours, that which is lost in dying may be restored in death—namely the natural complexion, which these cadavers have regained, so beautifully anew that they seem to live again and to sleep: all the limbs, which were so stiff in death, prove in these to be as flexible as in a living person.

Moreover, deceased bodies soon emit an unpleasant odor, yet these are pleasing of scent.

In this Anatomical treatment, it is not my intention to proceed systematically, as is my usual practice; for this should be considered something like a sampling of previous demonstrations, in which I will show different things, overlooked by others, as well as some new discoveries.

[Ruysch goes on to describe a dozen different Anatomical demonstrations he plans to undertake, five of which focus on various aspects of blood vessels—thus bolstering his claim that his observations predated those of Vieussens.]

I have decided to hold twice-weekly demonstrations, on Tuesdays & Thursdays, at three in the afternoon, starting on the first Tuesday in August, 1703, at the public Anatomical Theater, with a lecture in Dutch, or Latin, as necessary. All enthusiasts of Anatomy are invited to watch.

F. RUYSCH,

Professor, & President of the College of Surgeons.

Later, I began to publicize my preparations by means of a written account of my Anatomical Cabinets, with many appended figures taken from life, and there are XVI of these said Cabinets, V of which have already been published; now, since completing this task has demanded a great deal of time, & in the meanwhile, people come to me, nearly every single day, for an opportunity to inquire about one thing or another, or to view, to sketch, or to take notes for others, it is therefore no wonder that some of my discoveries have been published by others; nor has this happened only with matters that I had not yet disclosed, but even subjects I had already described, depicted, & published years ago, as I see has been done by the Illustrious Master Raymond Vieussens, a Physician at Montpelier, and by Master Hovius,[157] a Medical Doctor at Enkhuizen.

The first of them only recently composed the treatise, *A New System of the Blood Vessels in the Human Body*, in which I see that he has claimed for himself several of my discoveries some years after I described & published them with illustrations.

Among everything that I have discovered over the past 40 years, one thing truly stands out: namely, that the Cortical substance of the Brain is not glandular—as Anatomists had mistakenly reported, depicted, and indeed held as certain—but completely vascular.[158]

When, in 1698, I discovered this new method of preparation, the Most Illustrious Michael Ernst Etmüller, a Man no less learned than zealous in the study of Anatomy, would often visit to look at my Anatomical curiosities, among which there was none he found more astonishing than the "vascular-eiderdown" Cortex of the Brain (so it may justly be called, because the cortex of the brain is nothing but vessels, as delicate as down feathers); indeed, even as he gazed upon the Brain Cortex, thunderstruck, he urged me to publish so important a discovery; accordingly, he wrote me a letter inquiring into this Anatomical matter in 1699, which letter, moreover, is on offer for sale by the Book Seller, Johannes Wolters, of Amsterdam, along with my response, with appended figures, drawn from life.

After the said Most Illustrious Etmüller reviewed in that letter everything the experts had written on the glandular theory of the Cortex of the Brain, & showed that everyone, including the Most Illustrious Master Vieussens, agreed on this, he admonished me with these words: *Thus, Illustrious Sir, I ask, nay I insist, that you will not disdain to lay bare the true texture of the Cerebral Cortex, reveal its appearance, and define its function.* That the Most Illustrious Vieussens had not read my writings (so he said), or that no one informed him of what I had published at the time does not much surprise me; what does, however, very greatly surprise is that only after I disclosed my findings did the Illustrious gentleman reject what he had written before, and now follow in my footsteps.

[...]

Explanation of the Figures of the Sixth Anatomical Cabinet.

Table I [see image on following page]

Fig. 1. Represents a Human Skele]on, the size of a pinky finger, whose right hand holds an unfertilized human Egg (marked Letter A). The left hand holds three strands of hair, with three human eggs of different sizes.

Fig. 2. Shows a human Skeleton, slightly larger than the previous, from whose right hand hangs a fertilized human Egg, discharged from the womb after growing somewhat, & unopened, so that I have not been able to see its contents.

A. Rudiments of membrane.

B. Rudiments of blood vessels: constituting the early Placenta.

Fig. 3. A human Skeleton, the size of an index finger; from a strand of hair in its left hand hangs a fertilized Egg, almost the same size as the other one, and which I have opened, but in whose cavity I found nothing solid.

A. The exterior of the Egg, full of Vessels.

B. The hollow of the Egg.

Fig. 4. A rudimentary Placenta, which remained in the womb for some time after the embryo was discharged, whence it hardened, & lost its natural constitution.

Fig. 5. A rudimentary Placenta from a tiny human embryo.

A. A clump of Vessels.

B. A blood clot, which coheres more firmly than the aforesaid clump.

Table II [see image on page 179] represents V objects, the first of which is:

Fig. 1. A human Skeleton of about three months, from whose hands hang the rudiments of two human embryos, on threads of hair; the first of them, on the skeleton's right, is about the size of an anise seed. On the left, from a braid of hair, hangs the rudiments of a human placenta, to which is attached, by a rudimentary umbilical cord, the rudiments of a human embryo, the size of a louse or a lettuce seed.

A Human Fetus: Explanation of the first Rudiment

A. A rudimentary Umbilical Cord, hardly less stout than the rudimentary embryo, which is something I have happened to see several times.

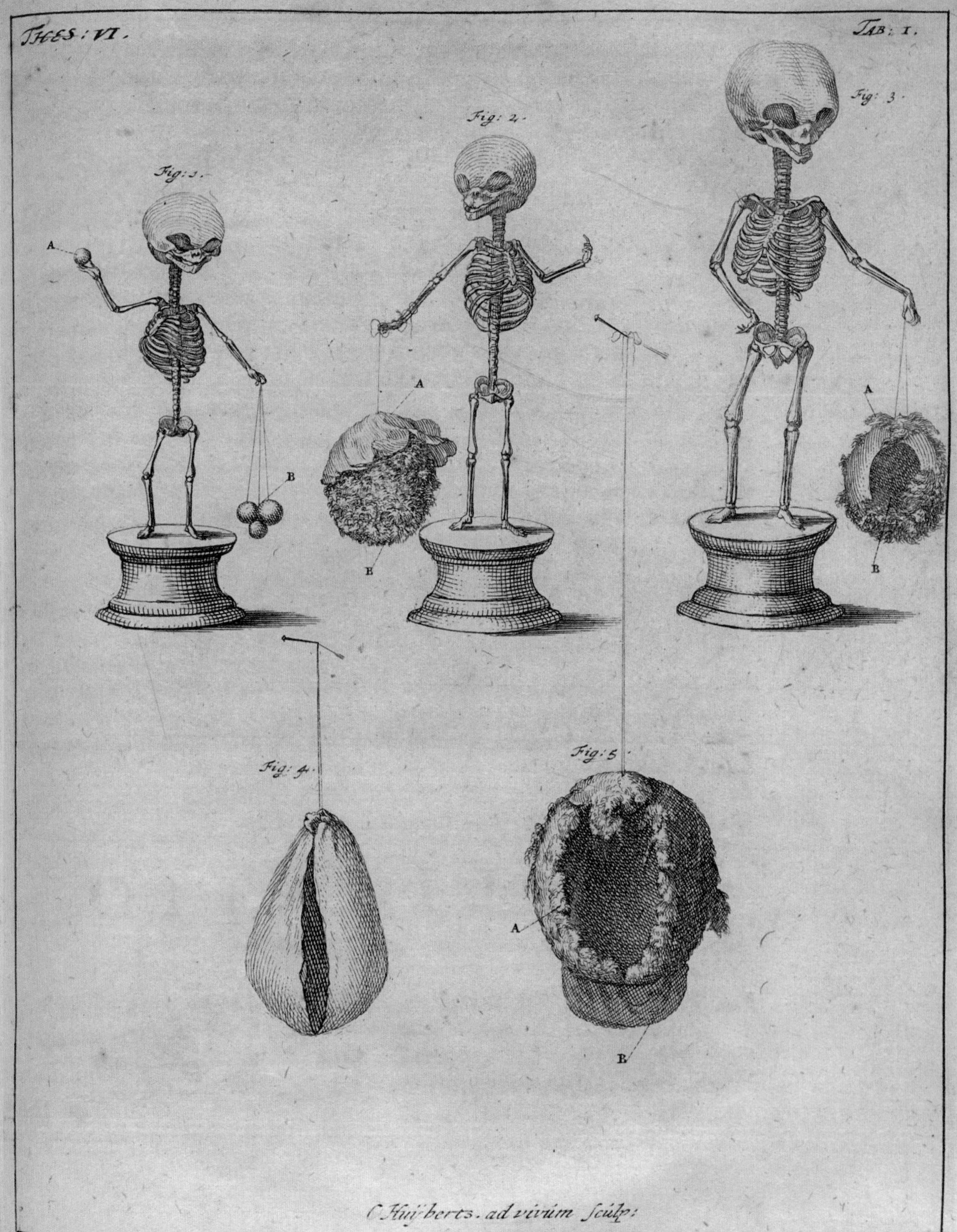

THES: VI.
TAB: I.
Fig: 1.
Fig: 2.
Fig: 3.
Fig: 4.
Fig: 5
A
B
A
B
A
B
A
B
C Huijberts. ad vivum Sculp:

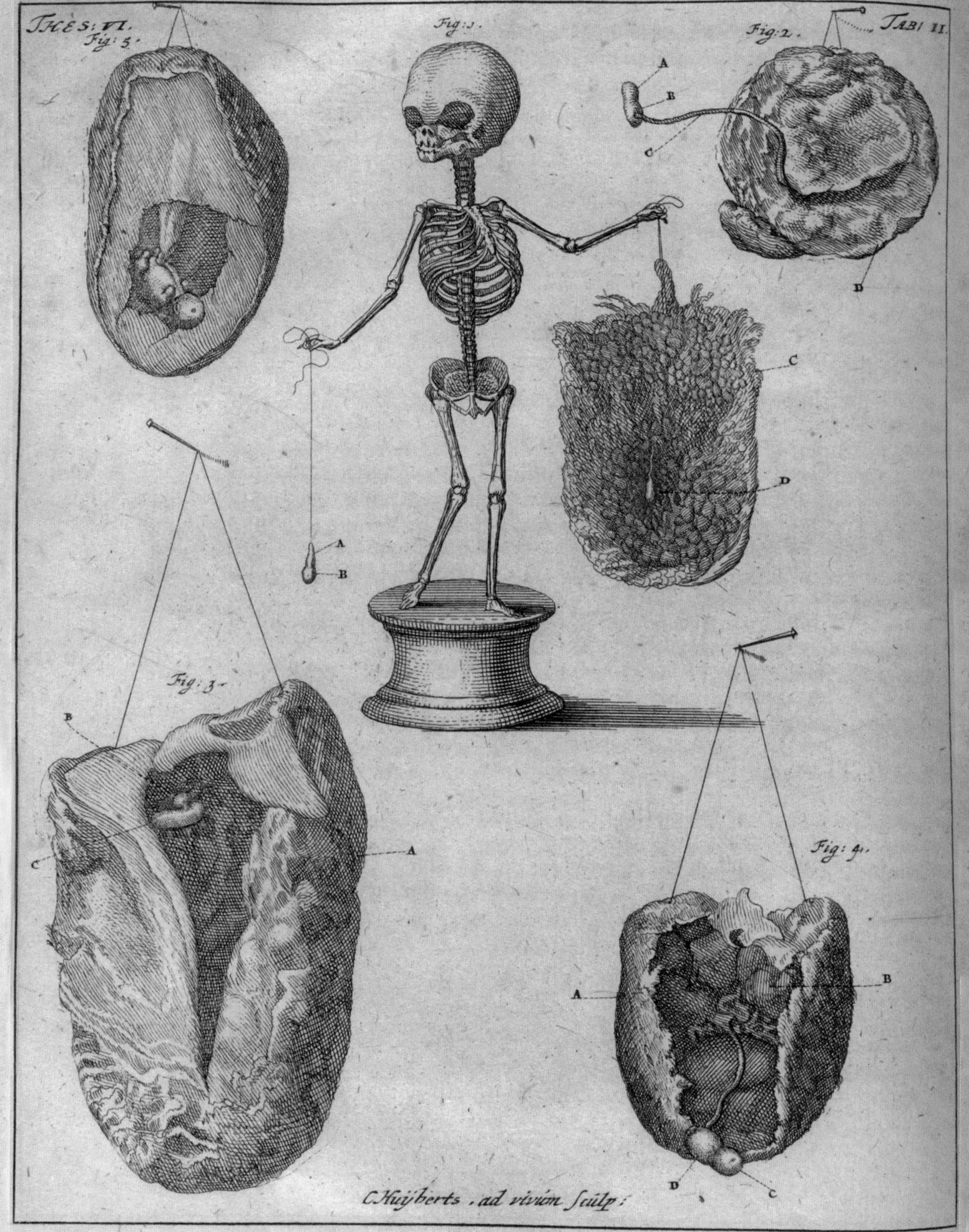
THES: VI.
Fig: 5.
Fig: 1.
Fig: 2.
TAB: II.
A
B
C
D
C
D
A
B
Fig: 3.
B
C
A
Fig: 4.
A
B
D
C
C. Huijberts, ad vivum Sculp:

B. The rudiments of an Embryo, the size of an anise seed.

Explanation of the second Rudiment

C. The rudiments of a Placenta.

D. The rudiments of an Embryo, the size of a louse or a lettuce seed.

Fig. 2. Shows a human Embryo, the size of a husked grain of barley, attached by its umbilical cord to the Placenta.

A. The Head of the Embryo.

B. The Body, not yet furnished with limbs.

C. The umbilical Cord.

D. The Placenta.

Fig. 3. Displays a human embryo, about the same size as the previous one.

A. The Placenta of an Embryo, of truly prodigious size for such an embryo, but, truth to tell, this is due to a blood clot that adheres so firmly to the actual Placenta that I could not separate them without tearing the placenta.

B. A rudimentary umbilical cord, which has degenerated into a water cyst.

C. An Embryo, whose head had not yet differentiated from the body.

Fig. 4. Shows a human embryo, a little larger than the previous, whose head is now distinct from its body, and stubby, bump-like limbs are coming into view.

A. The outer face of the Placenta.

B. The same, from the interior.

C. The head of the embryo.

D. The body of the same.

Fig. 5. Shows an embryo, a finger's breadth in size; not only can the head be distinguished from its body, but the rudimentary limbs are also jutting out like large bumps; & note that the rudimentary umbilical Cord is nearly as thick as the embryo itself, which, as I have said, I recall seeing many times. Could this be from a diseased condition of the Cord?

Table III. [see image on facing page] Explanation, in which VI figures[159] are found, of which

1. Shows a human skeleton, of about four months, from whose right hand hangs an Embryo on a strand of hair, slightly larger than the preceding one, whose limbs have largely developed, but whose fingers and toes do not yet appear to sight. This said embryo is marked as Fig. 2, and its umbilical cord (denoted with the letter A) is exceedingly stout. The embryo hanging from the left hand is a human fetus (Fig. 3), a bit larger than the other. In this

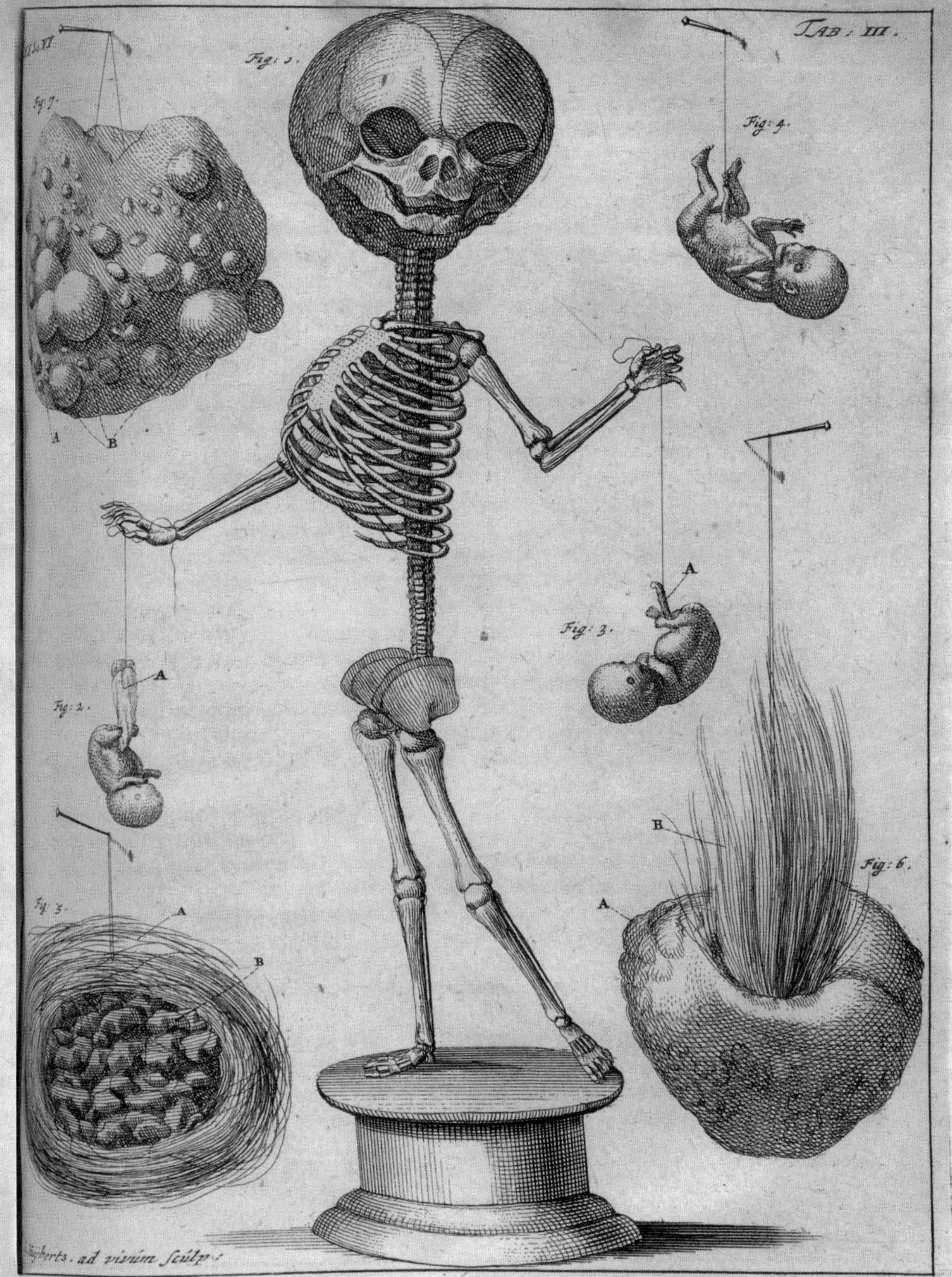
Tab: III.
Fig: 1.
Fig: 7.
A
B
Fig: 4.
Fig: 2.
A
Fig: 3.
A
Fig: 5.
A
B
B
Fig: 6.
A
ad vivum Sculp:

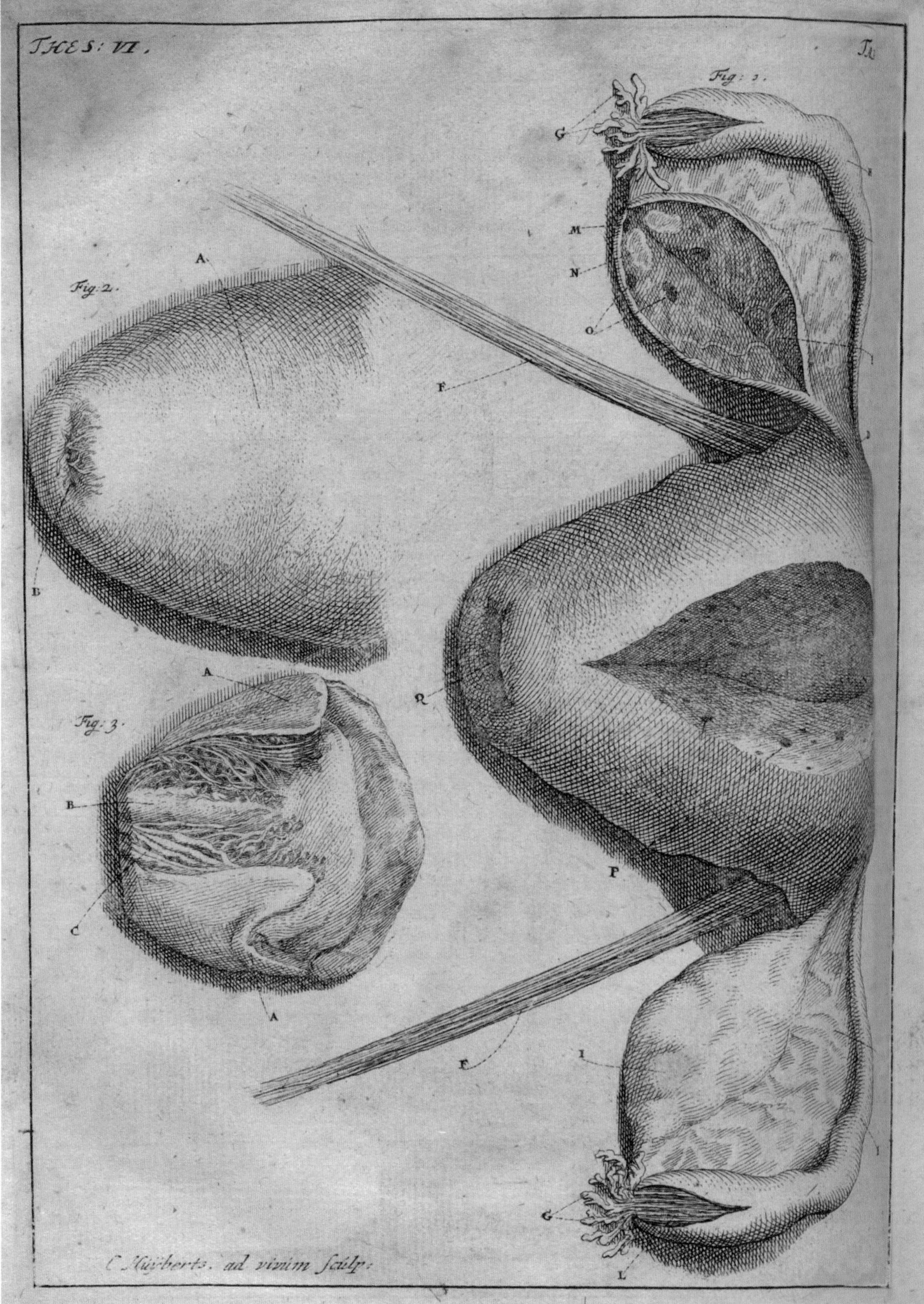
THES: VI.
Tab.
Fig: 1.
Fig. 2.
Fig: 3.
C. Huÿberts. ad vivum sculp.

fetus, all the digits can be Seen, not only of the hand, but of the feet as well, and the umbilical cord is thinner by far; so this may be considered a freak of nature.

It should be noted that I take this last example as a fetus, for the reason that all its members, including its fingers and toes, have been developed, and the umbilical cord is much thinner than in the preceding one.

Fig. 4. A human Fetus, hanging from a strand of hair, larger than the preceding fetus, and whose umbilical cord is just slightly thicker.

Fig. 5. A hairy Atheroma that was taken from between the neck muscles of a cow, & made to resemble the nest of the *aviculae Ourussiae*, commonly called *Colubritje*.[160]

A. Hairs arranged in a circle.

B. The Porridge-like substance found inside the tumor.

Fig. 6. An Atheroma taken from the same area, from whose cavity hairs emerge straight upward, and it represents a bowl of burning incense.

A. A hard & heterogeneous substance.

B. Hairs.

Fig. 7. This shows a human Placenta, which remained inside the womb for some time after the fetus was expelled; the interior surface of this is covered with hard bumps.

A. The Placenta.

B. Bumps.

Table IV [see image on facing page] shows III figures, of which the first is the Womb of a woman, with child for a few days.

A. The body of the Womb.

B. The thickness of the Womb.

C. Its hollow.

D. The place where the egg ducts (Fallopian Tubes) enter the womb.

E. An Ovary, bisected & opened.

F. The *Ligamenta tiretia*, commonly called the round ligaments.

G. The fringes of the Tubes.

H. The Fallopian Tubes.

I. Eggs, bulging through the surface of the Ovaries.

K. The side ligaments of the Womb.

L. The hollows of the Tubes.

M. The edge of the opened ovary.

N. An Egg, fertilized & flattened into a large mass, whose contents seem to have curdled, perhaps from my fluid, in which the object is preserved.

O.O. Unfertilized eggs, cut crosswise, whose contents have not curdled, although they are held in the same fluid.

P. Severed blood vessels.

Q. The internal mouth of the womb.

Fig. 2. Is a Womb that was prolapsed from the body of the woman.

A. The foremost part of the womb, covered by the Vagina, whence the Womb, although prolapsed, cannot immediately be seen; this holds true for every displaced womb, except for during birth.

B. The internal mouth of the womb, in which are seen the various furrows of the interior cervix, which here come into view by pressing, & by these efforts, women who are troubled with this said affliction are treated.

Fig. 3. The internal mouth of the womb, with the internal cervix of the womb, which has been opened lengthwise, so that the interior may be seen.

A. The side of the womb Mouth, cut through the middle.

B. The hollow of the cervix.

C. Furrows.

Table V [see image on facing page] contains VI Figures,[161] the first of which, marked as Fig. 1, is the Womb of a woman who was pregnant for a few hours, &, being caught in adultery by her husband, was killed a few hours later; Virile seed fills not just the hollow of the womb, but even all the way to the Fallopian Tubes.

A. The body of the womb.

B. The hollow of the womb.

C. Virile seed, somewhat hardened from the liquid in which it is kept.

D. The internal mouth of the womb, in which a little seed can still be found.

E. Ligamenta rotunda.

F. Fallopian Tubes.

G. Ovaries, which were once called the "female Testicles."

Fig. 2. A sample of the Ileum of the bowel, whose cellular lining has been inflated, dried, & cut across.

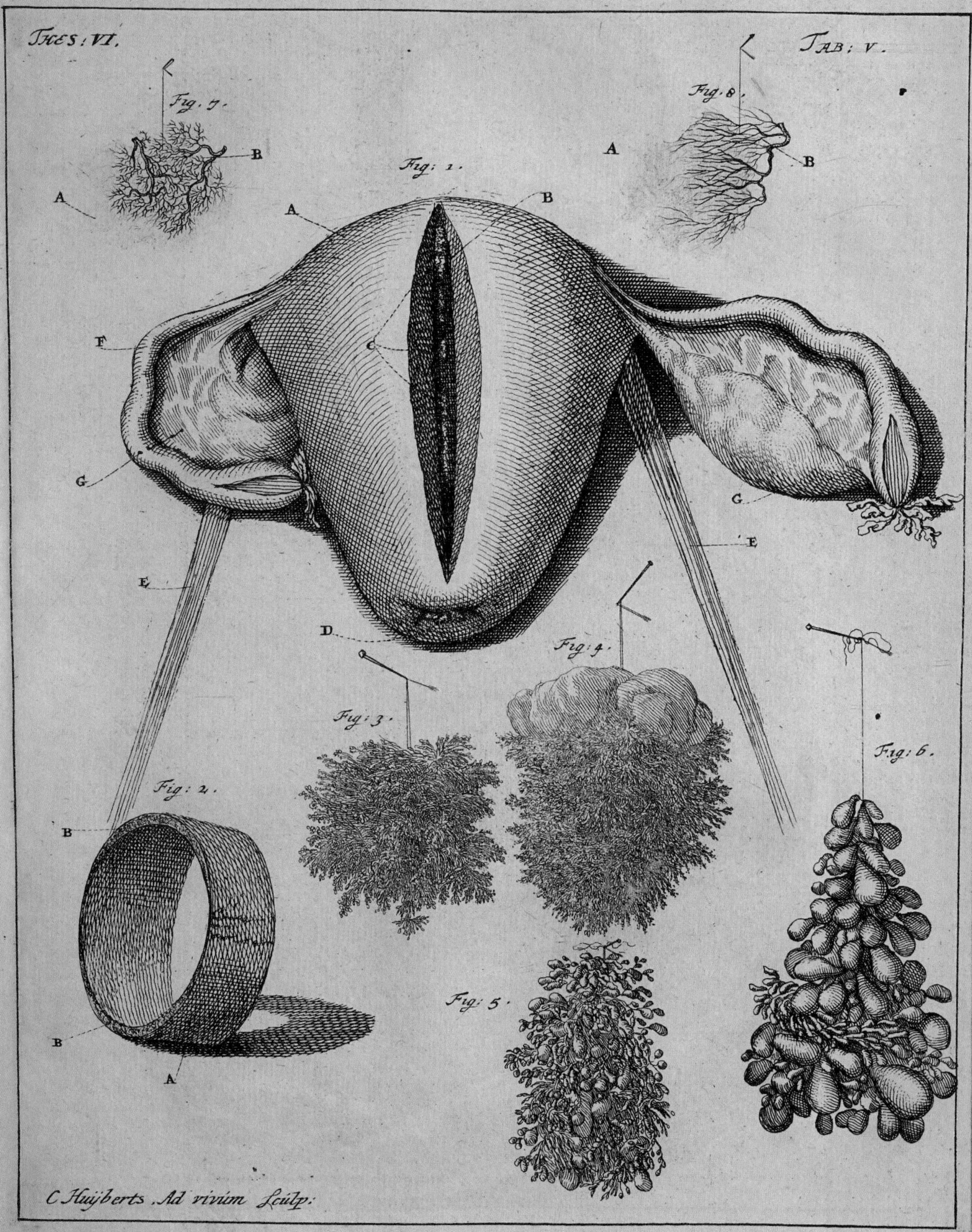
Thes: VI.
Tab: V.
Fig. 7.
Fig. 8
Fig: 1.
Fig: 2.
Fig: 3.
Fig: 4.
Fig: 5.
Fig: 6.
A
B
C
D
E
F
G
C Huijberts. Ad vivum Sculp:

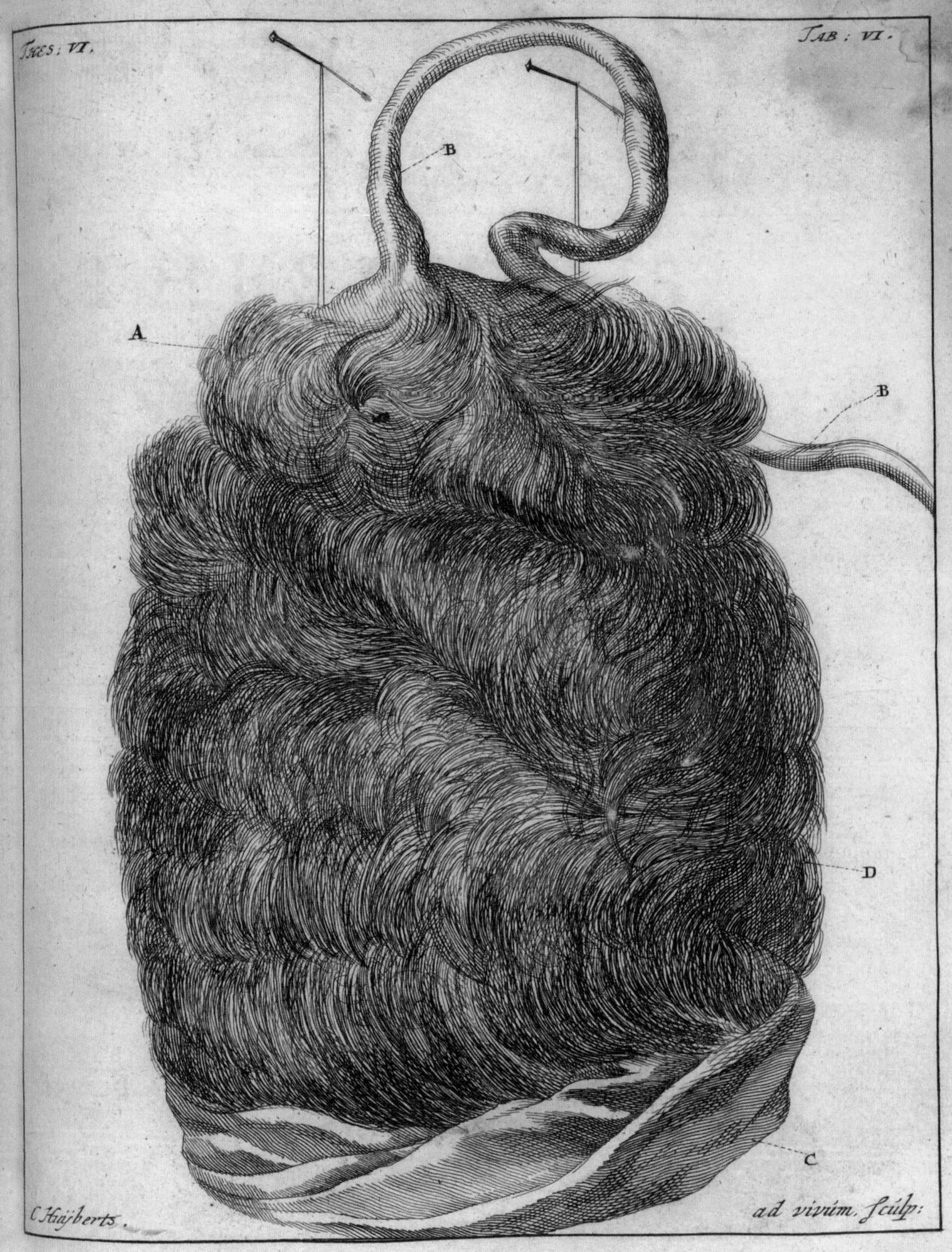
TAB : VI.
TAB : VI.
B
A
B
D
C
C Huijberts.
ad vivum. sculp:

A. The said piece of bowel.

B. The cellular lining.

Fig. 3 & 4. Shows pieces of human uterine Placenta, which remained in the womb for a few days after the fetus was expelled, whence the ends of the blood vessels in various places started to assume the form of water cysts.

Fig. 5. Displays part of a human placenta, in which the above-said can clearly be seen, with water cysts growing into a large mass.

Fig. 6. Indicates a piece of the same human placenta, which has devolved into so many water cysts. All these are preserved in my museum, as if they were just excreted from the body.

Table VI [see image on facing page] shows a certain preternatural body expelled from the womb of a cow, & endowed almost everywhere with short, dense, black, and white hairs, and with a very long, thin tail.

A. A hairy Body.

B. A little Tail.

C. A section without any hair.

D. Hairs.

Whoever may now have examined all the Embryos that I have represented here will readily object to the calculations behind my assertion: to wit, that it is difficult and troublesome to determine from the size of an embryo the time of its conception—especially since very few mothers can be found who can note the correct moment of impregnation. Moreover, Embryos have often lost their vital force, & nevertheless remained in the womb for some time, so I must hold that it is nearly impossible to determine aright age from the size of the Embryo. Concerning a human fetus of three, four, or five months, this can be decided with a little more certainty, since very often living ones come to light, & reach our hands in great number. And thus I have preferred to act more cautiously than those who incautiously assert that they can get this from an Embryo's size.

END.

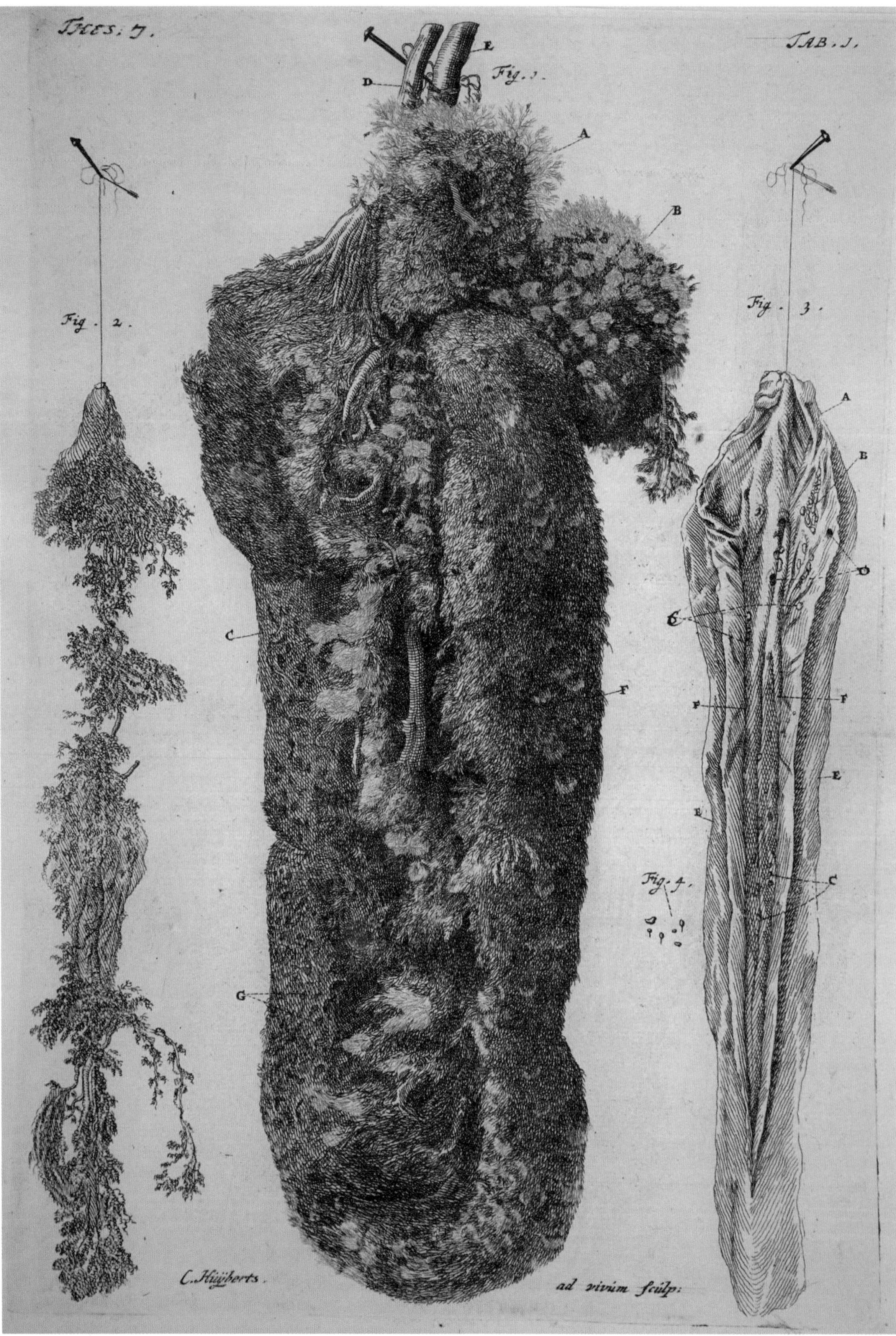
TAB. I.
Fig. 1.
Fig. 2.
Fig. 3.
Fig. 4.
C. Huijberts.
ad vivum sculp:

Frederik Ruysch,

Professor of Anatomy & Botany, and Member of the Imperial Academy of Natural History

The Seventh Anatomical Cabinet.

With Engravings.

Amsterdam
Johann Wolters
1707

EXPLANATION OF THE FIGURES OF CABINET VII

Table I. [see image on facing page]

Fig. 1. Shows the Spleen of a Gigantic girl.

A. The extremities of the blood vessels, entirely unfolded, as delicate as down or cotton wool; before this unfolding, however, their shape resembled that of Glands.

B. The said extremities, from the unfolded area.

C. An area on the surface of the Spleen that has not been unfolded.

D. The Splenic Artery.

E. The Splenic Vein.

F. Large holes, produced by pulling away the membrane surrounding the Spleen.

G. Small holes, produced from the same cause.

Fig. 2. Shows the Plexus Choriodes[162] from the above-said Gigantic girl, which is both very long and very delicate.

Fig. 3. Demonstrates a piece of human Dura mater, seen from the exterior face.

A. A piece of the Dura mater.

B. A throng of Millet-shaped little bodies, which are no larger than the head of a pin.

C. Individuals of those said bodies.

D. Holes, carved out in the longitudinal recess.

E. Part of the Dura mater, to the side of the longitudinal recess.

F. A longitudinal recess, opened lengthwise, & bent open somewhat, so that the hollow, & whatever it contains, may be seen.

Fig. 4. Shows the said bodies, removed from the fold, some of which have something like little feet.

Table II [see image on facing page]

Represents five objects, of which

Fig. 1. Shows the Forearm of an Infant (marked with the Letter A), in whose hand is a Fungus, that I discovered growing on a bit of flesh from a human corpse, and which resembles afterbirth.

B. Skin, resembling the Chorion.

C. The Fungus, like the Placenta.

Fig. 2. Shows a piece of human Brain, throughout whose surface, above the Pia mater, are disseminated Vessels of various shapes, resembling lymph Vessels, but which are nothing but the *Tunica Arachnoïdae*,[163] artificially raised, by blowing air through tubes. I decided to keep this Brain in my Cabinets, although I did not actually preserve it, since it was impossible for me to keep air in these said vessels. For more on this matter, see the Anatomical Program of 1706, committed to the press on the dog days, in which I said that, to that point, I had never encountered actual Lymph vessels either in the human Brain or in its membranes, & that I would be much indebted to whoever could show me even the least branch of an actual Lymph-duct, complete with its Valves.

A. A piece of Brain.

B. A pseudo-Lymph-vessel.

Fig. 3. Displays a piece of a Hydropsical Peritoneum, filled with water cysts.

A. A piece of Peritoneum.

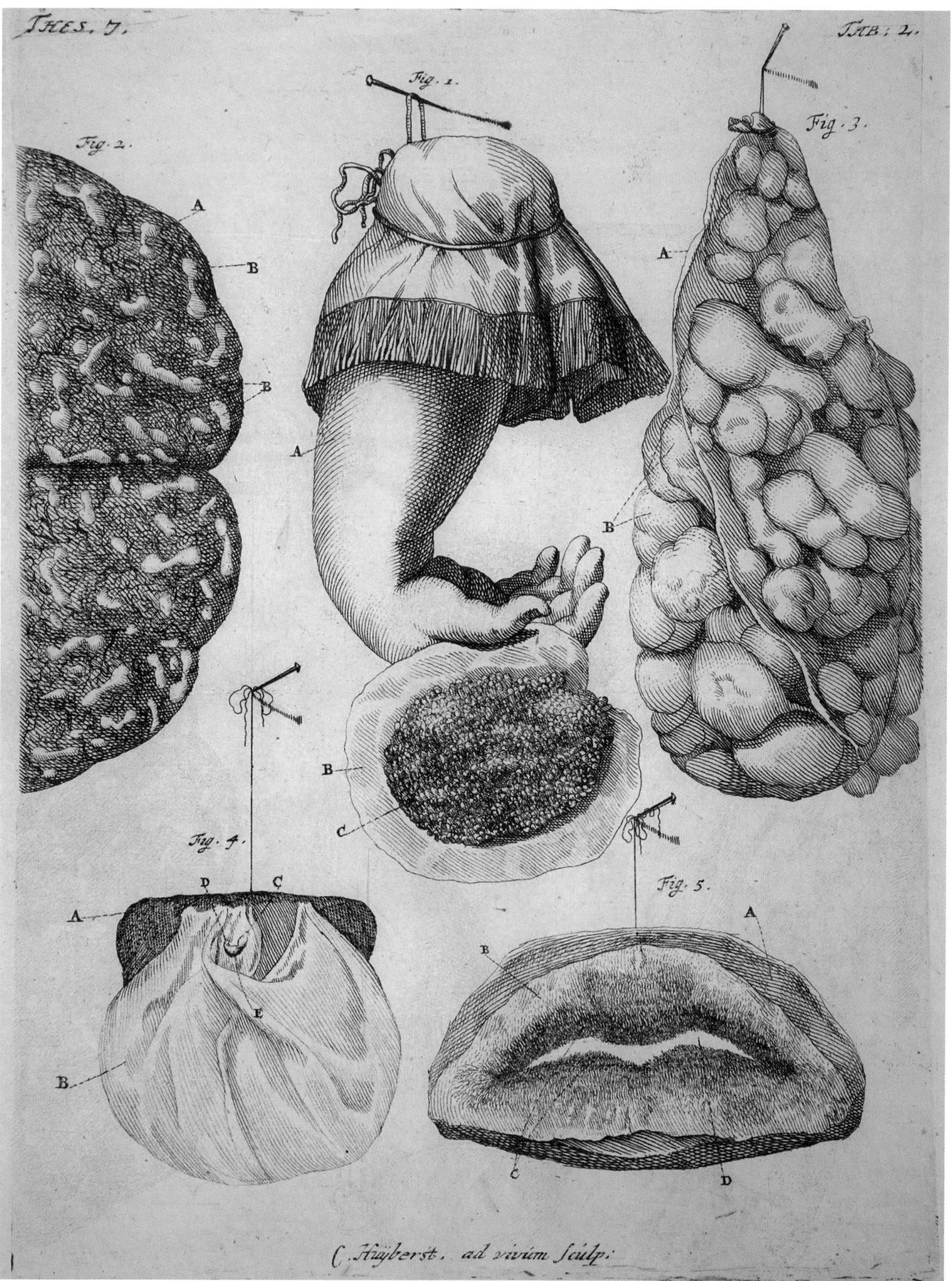
THES. 7.
TAB: 2.
Fig. 1.
Fig. 2.
Fig. 3.
Fig. 4.
Fig. 5.
A
B
B
A
B
C
A
B
D
C
A
B
E
A
B
C
D
C. Huyberst. ad vivum sculp.

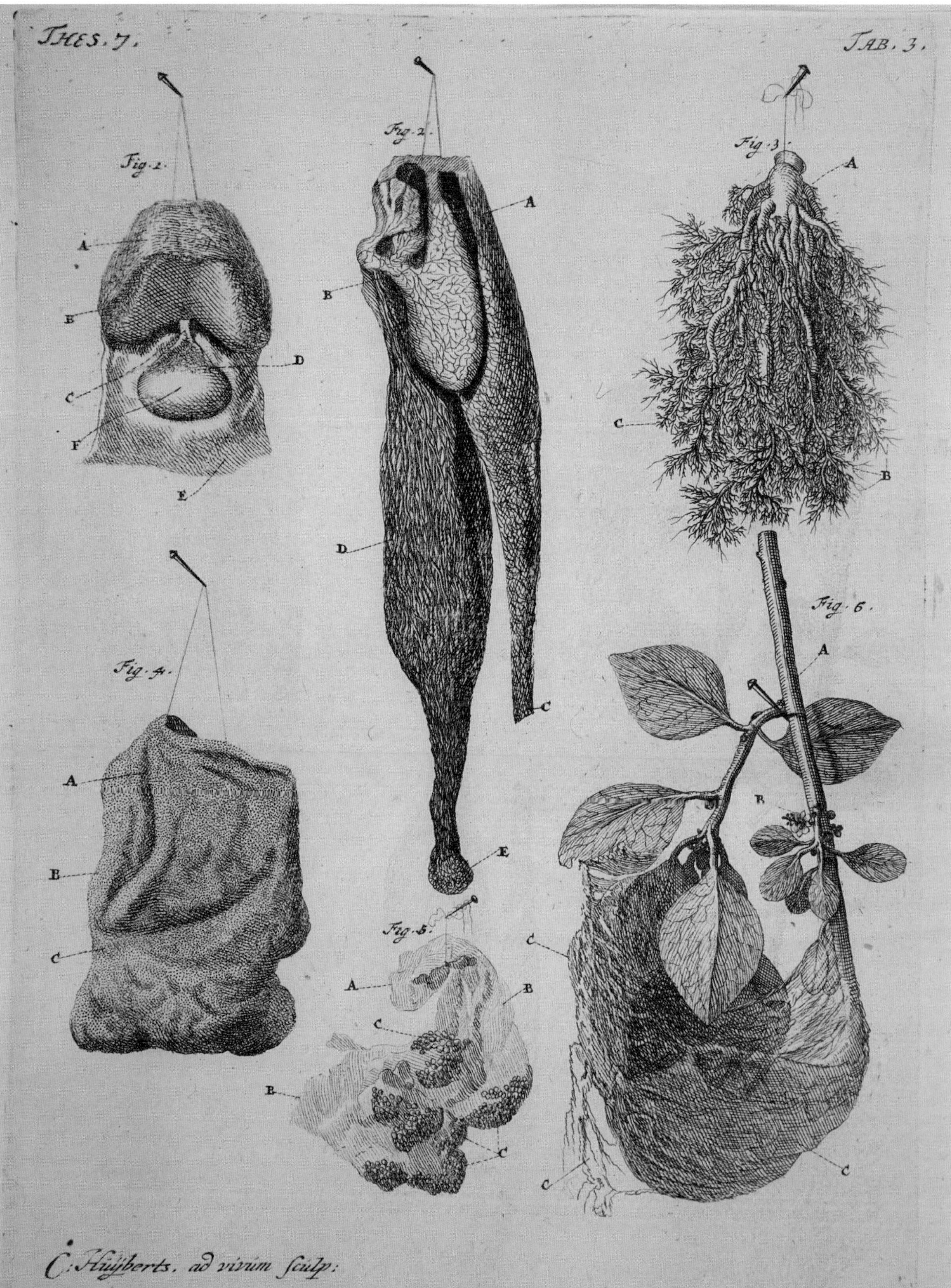
THES. 7.
TAB. 3.
Fig. 2.
Fig. 2.
Fig. 3
A
B
C
D
E
F
Fig. 4.
Fig. 5.
Fig. 6.
C: Huijberts. ad vivum sculp:

B. Water-cysts.

FIG. 4. EXPLANATION.

A. A mound of thick blood.

B. Membranes, covering the Embryo.

C. The rudiments of the umbilical Cord, almost surpassing the Embryo in thickness.

D. The head of the embryo, in which a single rudimentary eye can be seen.

E. The back of the Embryo.

Fig. 5. A pair of human Lips with part of the Cheeks.

A. The skin with the fat.

B. Epithelia[164] of the lips, & also the Cheeks.

C. Papillary nerves, like shaggy silk.

D. The opening of the Mouth.

Table III. [see image on facing page]

Explanation.

Fig. 1. Displays the lower Epiphysis of the Thigh bone, with the Kneecap attached, & the Teres ligament.

A. The lower epiphysis of the Thigh bone.

B. Arterioles distributed through the side of the Cartilage.

C. The Teres ligament, among whose functions is to connect the Kneecap to the lower Epiphysis of the Thigh bone.

D. Vessels, Arterioles in particular, dispersed throughout the substance of the cartilage.

E. Vestiges of the Tendons of the Muscles for flexing the Shin bone, from which the Kneecap firmly arises.

F. A hanging Kneecap.

Fig. 2. Shows the spongy bones of the Nostrils, from a sheep.

A. A piece of the septum of the nostrils, in its rather remote location, presented so that the spongy bones of the Nostrils should come clearly into view.

B. One of the upper spongy bones, through whose very envelope many vessels are dispersed, with a different crawl from the others that are distributed across the envelope of the lower spongy bones.

C. The interior part of the Nasal septum.

D. The lower spongy bone, across whose envelope are many Arterio-phlegmatic vessels, flowing in an entirely different crawl from those arteries in the upper spongy bone.

E. The lower part of the spongy bone, in which many mouths are found, through which a whey-like—or, better said, a whey-phlegmatic—juice flows spontaneously in the living, & can be squeezed through in the dead.

Fig. 3. Indicates a branch of Wind-pipe, filled to the tips with a red, wax-like substance. Here the great difference can be seen between this, & other figures published by others.

Fig. 4. A piece of a Child's Straight Gut,[165] seen from the inside, in which countless openings, or mouths, are seen.

Fig. 5. A degenerate *Plexus Choroides* from a sheep's brain.

A. A sagging section of membrane.

B. A section opened up.

C. Arterioles, constituting most of the plexus Choroides, & having degenerated into round balls.

Fig. 6.

A. A branch of the plant called "African Purslane."[166]

B. Little flowers, with five tiny petals, on a flowering stem.

C. An artificial membrane that I made from my own blood, after a blood-letting, with a single round of agitation. See the above-cited description.[167]

Table IV. [see image on facing page]

Fig. 1. A piece of Colon, with the blind-gut attached, along with the appendix, as well as a piece of Ileum from a Gigantic girl.

A. A piece of Colon.

B. A piece of Ileum.

C. The Appendix.

Fig. 2. A piece of Ileum from a deceased man, with an unnaturally large bulge.[168]

A. An unnaturally large bulge that resembles the head of a four-legged animal.

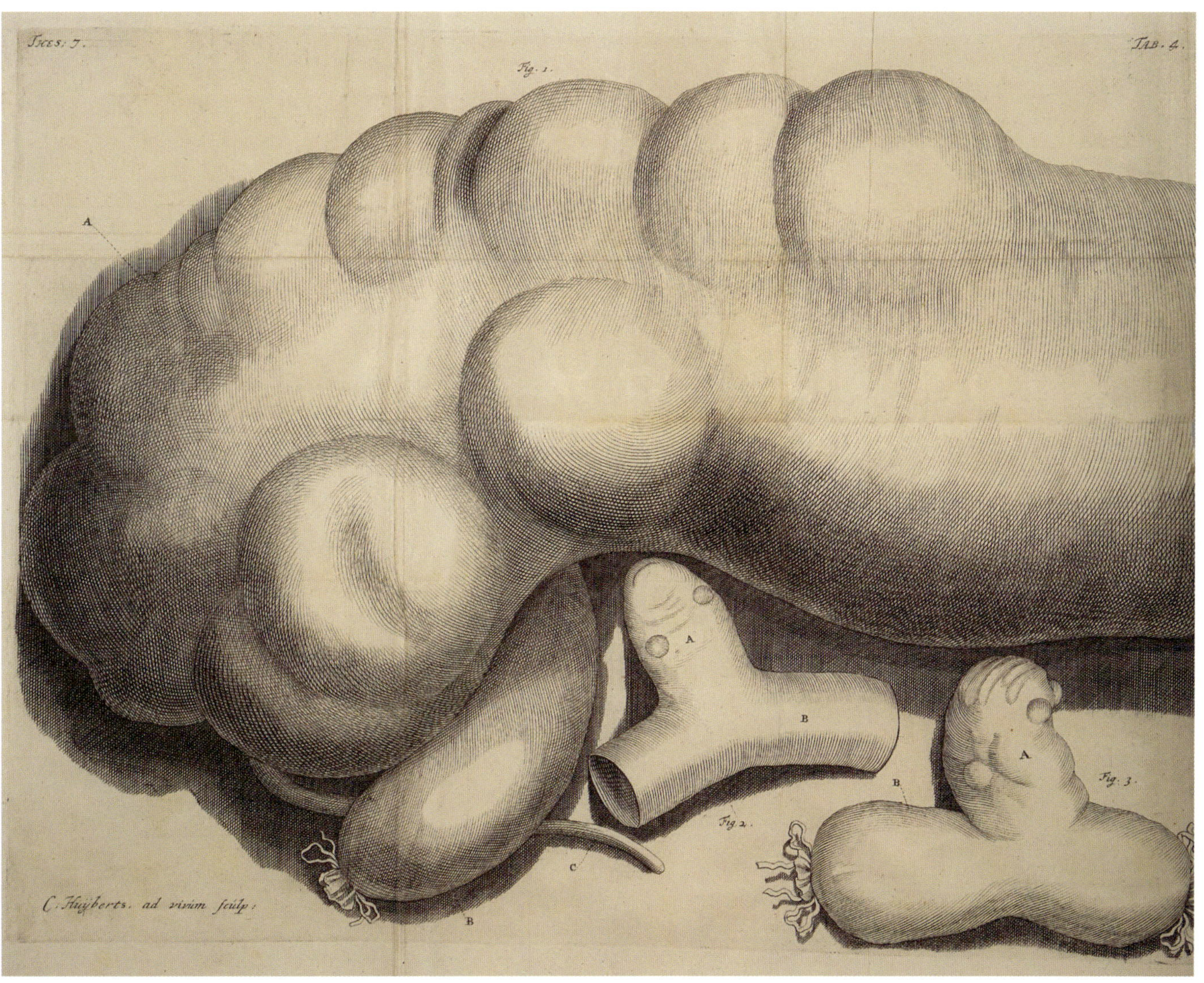

B. A piece of Ileum.

Fig. 3. Shows almost the same thing from a different man.

A. An abnormal bulge.

B. A piece of Ileum.

END.

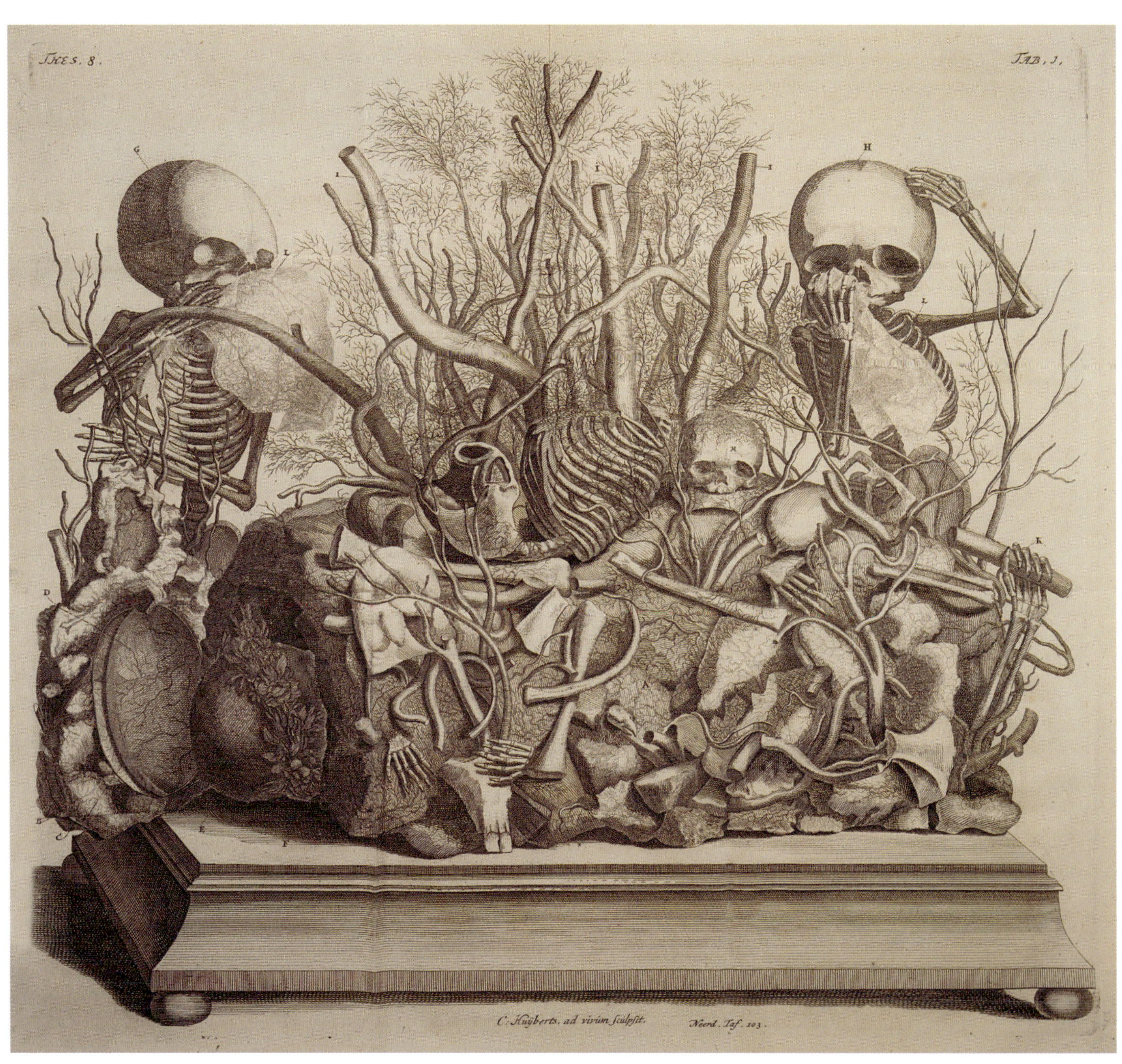
THES. 8.
TAB. I.
C: Huÿberts. ad vivum sculpsit.
Neerl. Taf. 103.

Frederik Ruysch,
Professor of Anatomy & Botany, and Member of the Imperial Academy of Natural History

The Eighth Anatomical Cabinet.

With Engravings.

Amsterdam
Johann Wolters,
1709.

EXPLANATION OF THE FIRST TABLE
[see image on facing page]

A. A Tomb, or sepulcher, containing a human fetus, of six months or thereabout, mummified by me about twenty years ago. This said Tomb is made of human bones, primarily from infants; & of Stones; membrane, sown throughout with countless arterioles; & Trunks & arteriole branches that resemble tree branches, or rather red coral.

B. The Lid of the Sepulcher, made from the upper part of an Infant's head bone; the exterior surface is covered with the same materials discussed above, and the bottom edge is ringed with dura mater.

C. A Sickle.

D. Dura mater, through which crawl many filled arteries.

E. The head of the immured[169] Fetus.

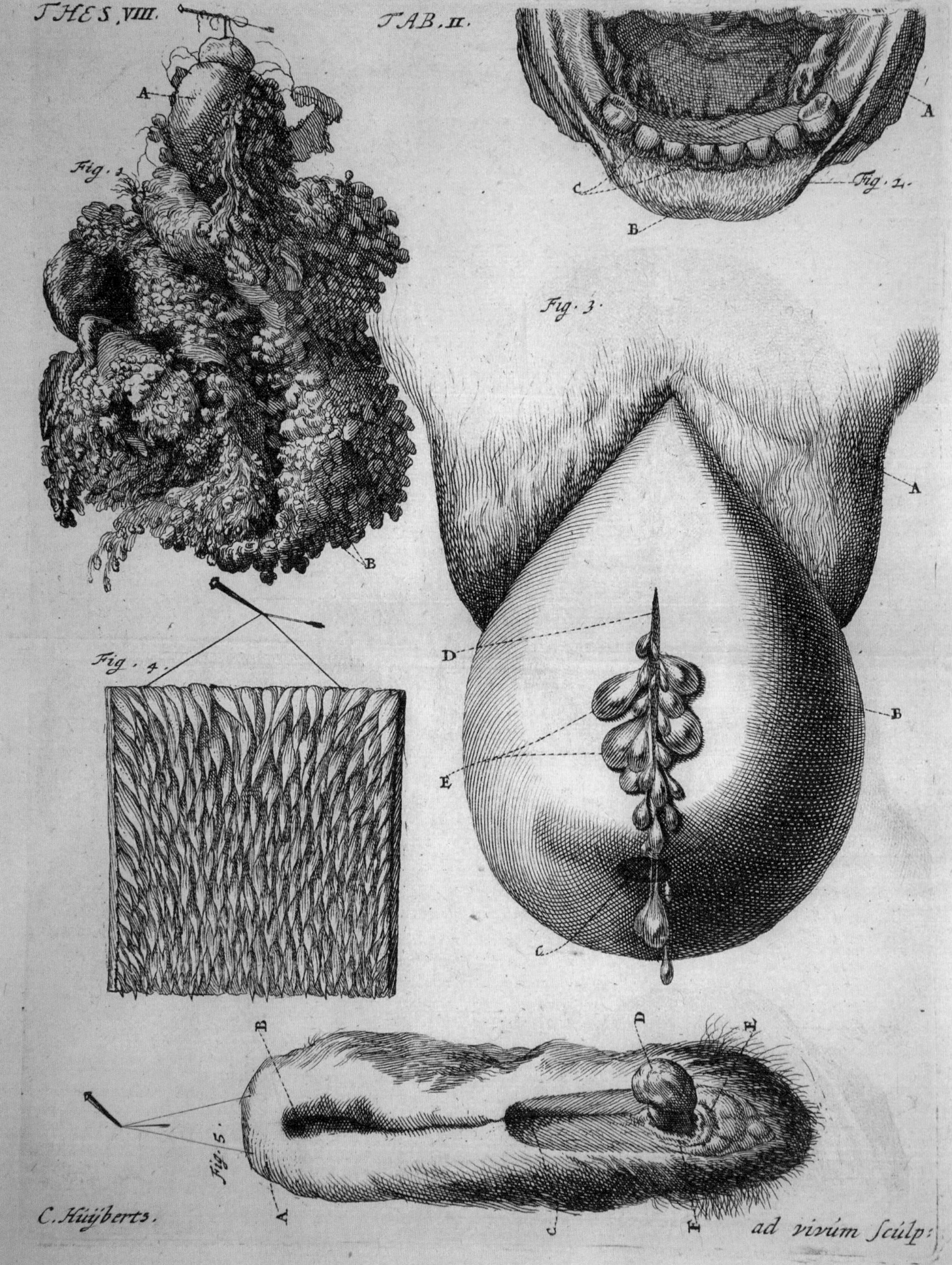
THES. VIII.
TAB. II.
A
Fig. 1
B
A
C
Fig. 2.
B
Fig. 3.
A
D
B
E
C
Fig. 4.
B
D
E
Fig. 5.
A
C
F
C. Huijberts.
ad vivum Sculp:

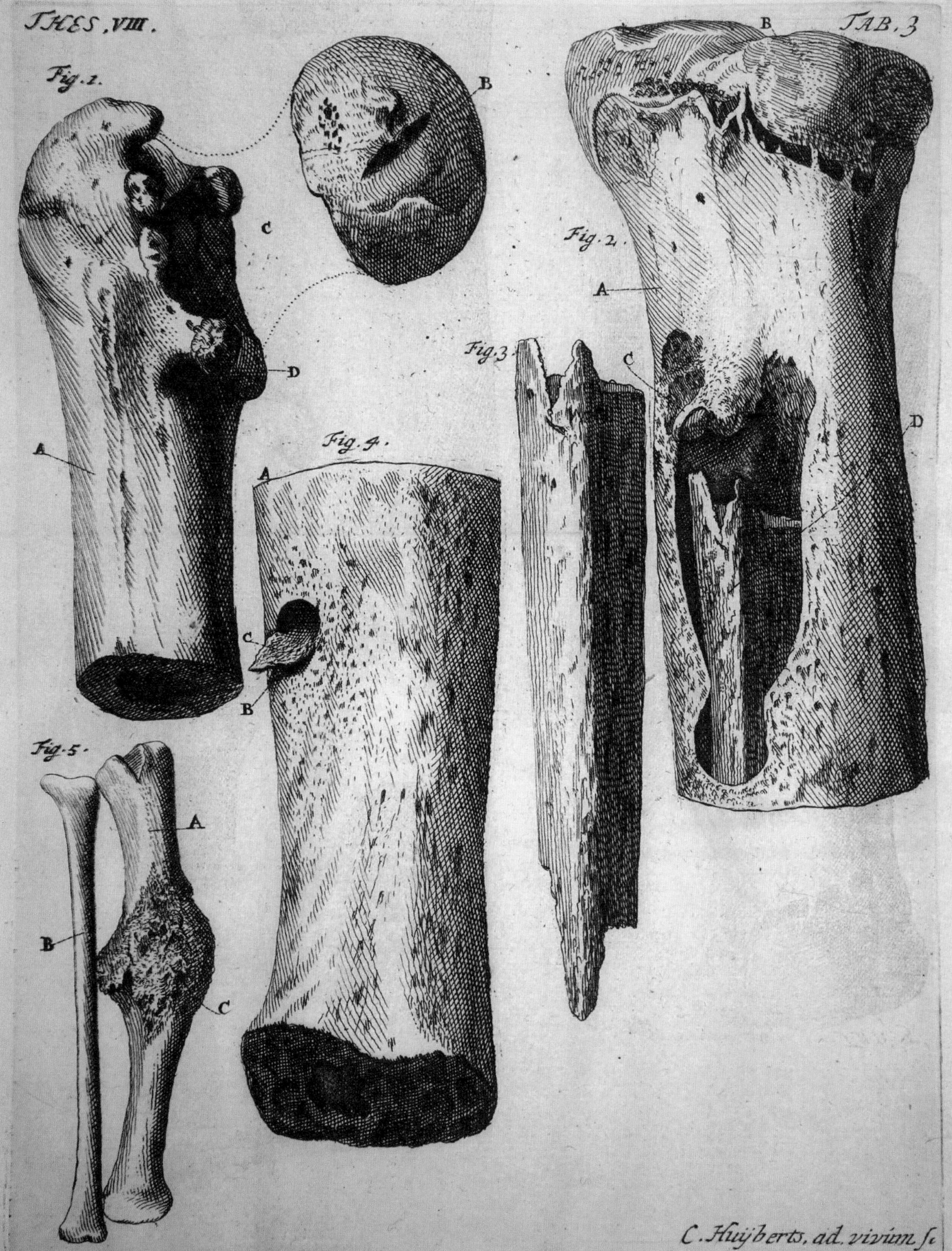
THES. VIII.
TAB. 3
Fig. 1.
Fig. 2.
Fig. 3
Fig. 4.
Fig. 5.
A
B
C
D
C. Huijberts, ad. vivum. fc.

F. A crown made of natural flowers, & placed upon the above head. The flowers include

G.H. The skeletons of twin premature fetuses, about six months old, behind the sepulcher, set on bended knees.

I.I. Arterial trunks, resembling saplings.

K. Bones of the Elbow & hand; the little Bones of the hands appear to clutch an arterial branch.

L.L. Two kerchiefs, made of membrane, boasting myriads of the most delicate arterioles.

M. The head of a premature fetal skeleton of about four months.

Table II [see image on page 198]

Contains five figures, the first of which

A. Denotes a portion of womb, whose interior surface has transformed into thousands of globular bodies, which deceptively resemble glands.

B. The said bodies.

FIG. 2. EXPLANATION.

A. Part of the lower jaw of a Child.

B. The lower lip, turned downward, & with the outer layer of skin removed, to allow the papillary nerves to come into view.

C. The papillary nerves of the gums.

D. The papillary nerves of the lower lip.

Fig. 3.

A. The privates of a woman.

B. The womb, hanging out of the body, & for the most part blocking the urinary bladder.

C. The internal mouth of the womb.

D. A lengthwise incision inflicted by a surgeon.

E. Droplets of urine, issuing from the Bladder through the incision.

FIG. 4. EXPLANATION.

This shows a portion of the Throat of a marine Tortoise, with its papillary nerves.

Fig. 5.

Shows the privates of a pseudo-hermaphrodite Sheep.

A. Skin near the anus.

B. The Anus.

C. The Vulva

D. The head of the clitoris, rising up in a great mass.

E. The Clitoral Hood.

F. The Cervix of the clitoris.

TABLE III

Explanation. *[see image on page 199]*

Fig. 1.

A. The upper part of a human thigh bone.

B. The upper Epiphysis.

C. The space for the neck, which neck is completely missing here, as would be expected after a fall.

D. The lesser Trochanter.

Fig. 2.

A. The upper part of a Shin Bone, eaten with decay.

B. The upper Epiphysis.

C. Decay.

D. A Fistula of the bone, which is actually the inner layer, separated from the exterior layer by brute force, & extracted by the hand of a surgeon.

Fig. 3.

The entire said Fistula.

Fig. 4.

A. The lower part of the same Shin bone.

B. More decay, from whose cavity a small bone (C) was growing.

Fig. 5.

Two little bones from a Duck's wing.

A. A little cubit bone.

B. A little ulna bone.

C. An imperfect Knitting of the bone, due to flaw in the outer shell.

END.

Frederik Ruysch

Professor of Anatomy & Botany, and Member of the Imperial Academy of Natural History

The Ninth Anatomical Cabinet

in which appears diverse information about the human body worthy of attention.

With Engravings

Amsterdam
Johann Wolters, 1714.

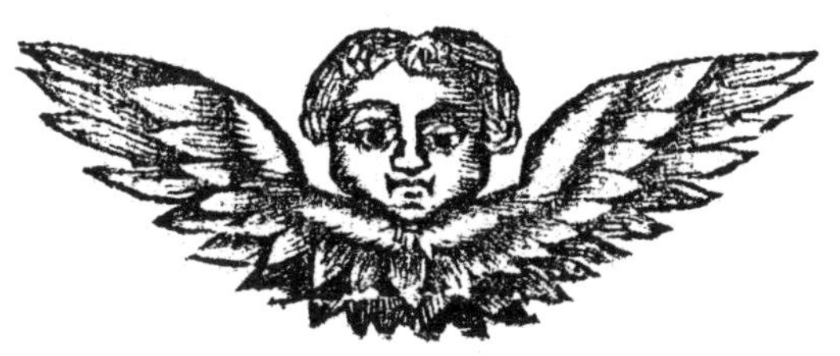

TO THE ILLUSTRIOUS
MOST DISTINGUISHED AND EXPERIENCED MASTER
ANDREAS VON GUNDELSHEIM,[170]

Royal Physician

To the Most Serene and Puissant

King of Prussia, and to his court

Consummate Botanist, &c.

May This Ninth Anatomical Cabinet Be
a Sacred Expression and token of Undying Friendship,

Frederik Ruysch.

To the most widely celebrated Gentleman
Frederik Ruysch.
To preserve may give less glory than to seek:[171]
Your art, Great man, on both accounts exceeds.
Immortal name of learning & technique,
No others' efforts will outlast your deeds.[172]

To the Ruyschian Cabinet.
The fabric of the human frame, whoe'er,
With 'stonished eyes, its works would fain to see,
Let him to RUYSCH'S Cabinets repair,
Constructed from his ingenuity
& skill of hand. This treasury unique,
None else could match, where'er you fain to seek.[173]
Heinrich Christian Krüger
Lüneburg[174]

Preface

TO THE KIND READER

According to the old saying: *He gives twice, who gives quickly.* Yet in this case, I confess, that hardly applies, for only now, long after the promised edition of the Ninth Cabinet, have I brought it to a conclusion.

I have no doubt, however, that you will excuse me, considering my multifarious occupations, both public and private, not the least of which would rank my medical practice, which is more than thriving, as well as my continuing investigations & examinations of human cadavers, which represent much of my activity, not only now, but in fact for more than fifty years, which is to say for the greatest part of my life, which has now reached its seventy-sixth year.

Consider, as well, that I have never been given a greater opportunity to examine cadavers than at this moment and for the past several years.

And, when it comes to opportunity, those who know me well know that I never spurn any I am given; rather, I attend to patients, night and day, and at break of dawn I am examining cadavers, so impartial judges of these matters will excuse me, if I have not committed this Ninth Cabinet more quickly to type.

Unless I am much mistaken, though, whosoever condescends to scroll attentively through this Ninth Cabinet will encounter things that they will find deserve the utmost astonishment.

For what mortal soul would have believed that, within the viscera, complete and untouched, with even the surrounding membrane still in place, a special technique could reveal—aye, & most clearly, too—the fleshy & juicy tips of the blood vessels; and show their proper location & natural shape, not just under a lens but even to the naked eye?

Indeed, I dare say, no one will easily calculate the value of these discoveries. That I am not vainly boasting, or saying anything other than the truth, is confirmed by all who view these discoveries with amazement at my home, on almost any given day; nor do I doubt that this is going to be regarded as a miracle by those who occupy themselves with examining cadavers; for they, no doubt, have often experienced for themselves how difficult it is to trace the extremities of the blood vessels all the way to their end; hence, the truth of their composition has lain hidden until now; and hence, it is not only the lower-tier Anatomists who have been led astray to faulty opinions, concluding that the viscera consist of supposed Glands, when, to the contrary, they are nothing but vessels with their pulpy extremities, which pulpy extremities form a continuous whole with their trunks, just as in a Cabbage, or Cauliflower, whose stalk tapers directly into florets, with nothing intervening; while, to the contrary, a gland is itself a part, & is enclosed in its own membrane.

Among these discoveries, I think none more worthy of admiration than the fact that the cortex of the Brain, whose consistency is softer than thick porridge, consists purely of the pulpy extremities of Arterioles, and not of supposed Glands.

What mortal would believe that, unless he had seen it at my home with his own eyes? Indeed, of everything that I have uncovered in the human body, this finding, I believe, may without offense be called my master stroke.

And thus no one would deny that it is disgraceful that Master Martin Lister, an Englishman, and champion of the theory of Glandular Viscera, in his treatise *On the Humors*, should argue that I put forward falsehoods, helter-skelter—an accusation that would be better turned back upon himself, who would judge these matters without ever seeing them for himself; certainly detractors of this sort, who judge matters about which they are completely ignorant, need hardly be engaged with.

However, he goes on, in various places in that said treatise, to attack me unjustly, saying that I believed & had written such & such, when I, in fact, had never stated nor written so; to wit, he related that I denied that any glands are produced in the human body; that I had not noticed that the Testicles are themselves glandular; that I spun a yarn that the Glands of the Brain are fatty, &c.—but why did he not add where or in what book I had written this? For whosoever would peruse my writing would find the contrary.

Be that as it may, I would not attack that polemicist simply because he perceived different things in various places than did I, but because he attempted to invalidate my writings, &, even in the face of better discoveries, still wanted to maintain the status of glands & to cling tenaciously to an outdated opinion, although it never occurred to him to view the new revelations & the actual preparations at my home.

With no less wonder have I read the treatise *A Journey Through England and Holland*, by Master Christian-Henri Erndl, a man of keen parts who writes on pages 83 & 84:

I have visited Master Ruysch quite frequently & by no means neglected those hours that he had set aside for demonstrations of Anatomy, or the showing of his Cabinets of Anatomical curiosities for the sake of visiting students.

An unbelievable abundance of Anatomical preparations & an astounding natural history Cabinet, which this most celebrated man keeps in his own house, so dumbfound the inquisitive viewer that to attempt to number these things is like plunging into a vast sea. Proof of this can be found in the VII volumes he had then published under the name The Cabinets, *not to mention the Anatomical Observations & Letters, that he has produced.*

Even so, I cannot but not remark upon something that stands out even among his rarities: the Mummy of an eight-year-old boy. The body is so nicely preserved, by means of Ruysch's embalming fluid & technique, that both the color & the consistency of the skin & the Musculature appear natural & alive to sight & to touch; each part & limb of this Mummy is not excessively stiff, but soft in constitution, as though still flush & hale with blood & vital humors.

This artistry, with which this Illustrious Man has prepared this Mummy, is to be marveled at, & is clearly something unheard of, surpassing all belief, unless one should see it oneself, & granted that the preparation of this cadaver might be assessed differently by his rivals, still there is no one who could produce or demonstrate anything comparable to this.

Dr. Rauw[175] *often said about this prodigy, when we asked for his opinion, that there are wicked deceptions, and slights of hand with that boy; that he removed the Scarf Skin*[176] *from the wretched boy, so that the true Skin, with its Muscles still full of blood, should recapture the natural color of a body, so to speak; others would talk about a vacuum, in which he kept the little corpse protected, except for the duration of the demonstrations.*

But all that crystal-ball gazing carries little weight; show me a parallel case & I would most gladly yield to your opinion.

How, pray, could the Illustrious Master J. Jakob Rau say that, since I displayed that very cadaver to the public, not once but several times—and He himself was there. GOD willing, he will even be at my next scheduled public Dissection, wherein I have decided to display once more that said cadaver, now preserved by me for 19 years, & and still manifesting the appearance of life.

Yet, if anyone would like to accuse me of "wicked deceptions," what would that be but pure calumny & envy? And, indeed, on the subject of the scarf skin, what does it matter if I had removed it—which is also altogether far from the truth: but, if I *had* removed it, would that be evidence I had committed wicked deceptions? or that, on this account, I had made a wretch of the boy?

I am often compelled to remove the scarf skin when I want to show the Papillary nerves; & whether the muscles are still filled with blood, I leave it to others to judge, but even conceding that point, this is still no trickery.

And so, if Professor Rau has often spoken as Master Erndl says, then he has misjudged these matters, &, therefore, I should answer: if he had learned better, he would have acted better.

Deception is something to be detested at all times, & it particularly dishonors a man of my advanced age, who stands at death's door; if Master Erndl had not committed this to print, I would have said not a word but simply spun wool; however, since this has now been sent to the press, I have been compelled to defend myself.

Having spoken in my own defense, let me turn to other matters—to wit, in this very Cabinet IX there also appears a four-legged animal like a dog, supplied with IV paws, a head, tail, mouth, tongue & something like an umbilical cord & preserved in my embalming fluid, along with the sac (instead of afterbirth) in which it was contained.

With that, farewell, Kind Reader, & any typographical errors that have crept in, I ask that you may obligingly correct.

THE EXPLANATION OF THE TABLES OF THE NINTH CABINET

Table I [see image on facing page]

Fig. I. Shows the upper part, or head, of a Thigh bone.

A. The Epiphysis of the said bone.

B. The thick, tough Ligaments, which are in the place of the missing neck, and connect the Epiphysis, or head of the bone, with the greater Trochanter: both the ligaments emerge not from the exterior but from the interior surface.

C. Marks the point of weakness in the neck of the Thigh bone, which has entirely broken away.

Fig. II. Indicates the shin of a human fetus, hanging from an Atheroma in the uterine Placenta.

A.A. Is the umbilical Cord.

B.B. An Atheroma that grew from the middle of the placenta. According to an old superstition, such a Placenta would be consigned to the fire.

C. The Shin.

D. The bottom of a Foot, with three toes.

Fig. III. Shows a cross section of a Leg.

A. Skin.

B.B. Fat, below which no Muscle is evident.

C. A circular section of Thigh bone.

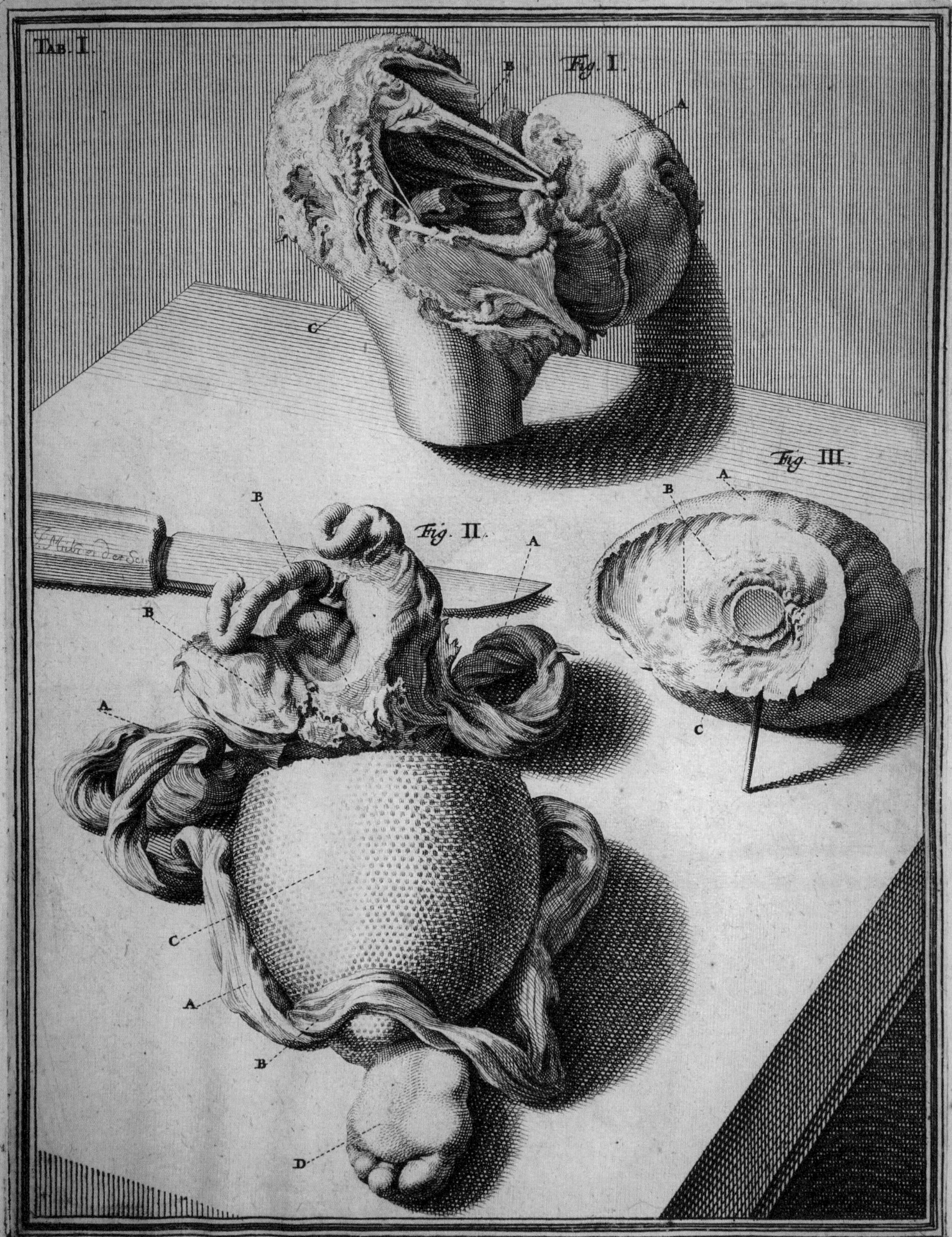

Tab. I.
Fig. I.
B
A
C
Fig. III.
A
B
C
B
Fig. II.
A
B
A
C
A
B
D

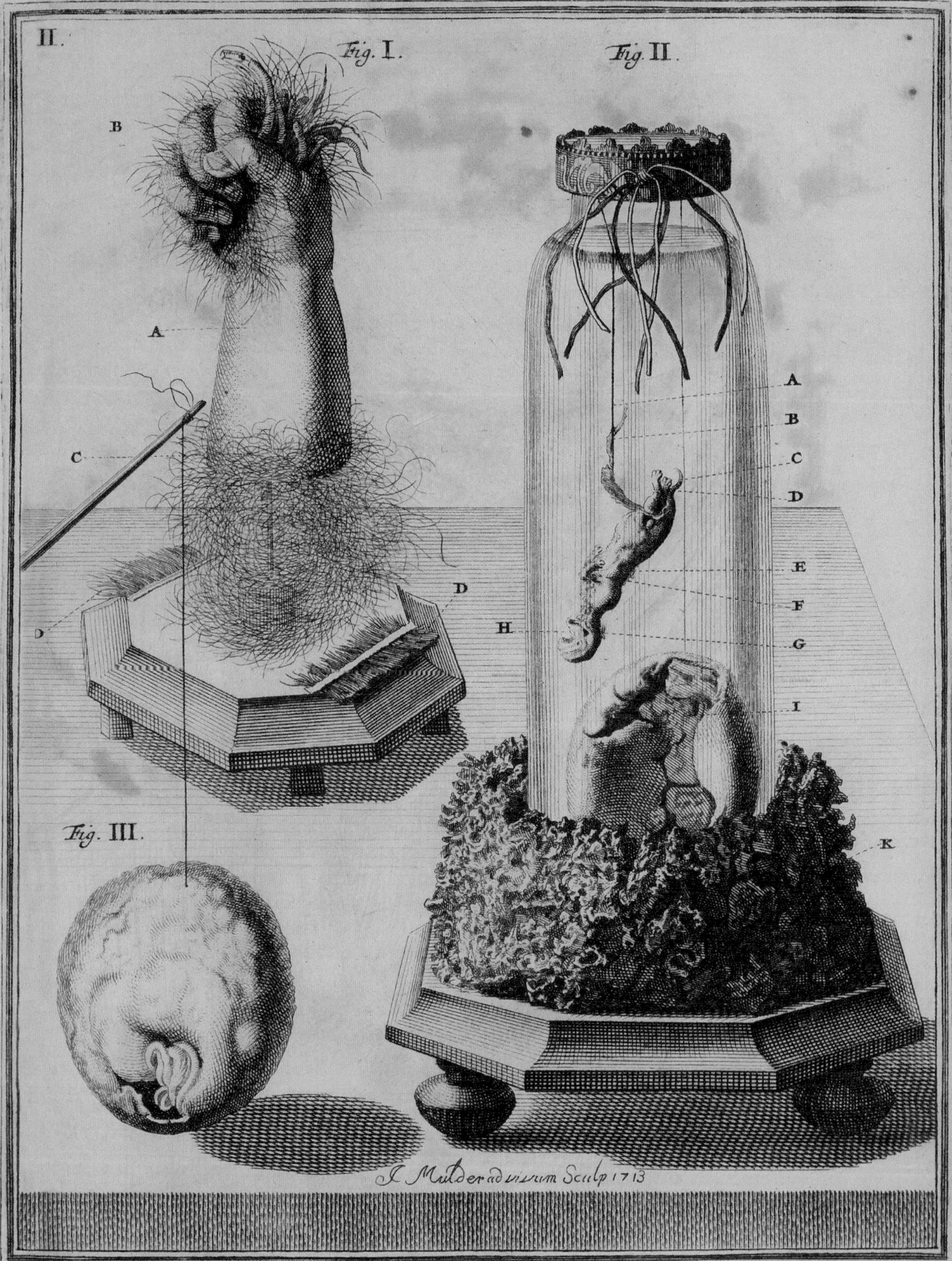
II.
Fig. I.
Fig. II.
B
A
C
D
D
A
B
C
D
E
F
H
G
I
Fig. III.
K
J. Mulder ad vivum Sculp 1713

Table II [see image on facing page]

Fig. I. Shows part of the forearm of a human fetus, hardened almost to stone by embalming, whose fist clutches a bundle of hair covered with a stony substance, resembling hornwort.[177]

A. Part of the Forearm.

B. Hair.

C. Hair that I found in an atheroma.

D. Hair that spontaneously fell from the head, as though suddenly cut with scissors.

Fig. II. A Jar containing a little four-legged Creature, like a Puppy, ejected by vomiting.

A. A hair, from which the creature hangs.

B. Something comparable an umbilical Cord.

C. Two rear paws.

D. A tail.

E. A partially exposed Shoulder Blade, the result of people incessantly touching it.

F. Forepaws.

G. The Head.

H. An open Mouth, in which a tongue can be seen most clearly, but which is not very well represented here.

I. The Sac that held the creature, seen from the side from which it burst forth.

K. A Stone, in the hollow of which the jar has been placed.

Fig. III. Shows the figure of the said sac, seen from below.

Table III [see image on following page]

Fig. I. Shows a jar full of liquid, in which is kept the forearm of a human fetus, whose hand contains a *lamella*, or slice from a human testicle of prodigious size, that has turned into cartilage: while I have cut the entire testicle into *lamellae,* I have represented here one of the smallest.

A. The upper part of the forearm, covered and bound in linen.

B. The fingers of the hand.

C. The slice of Testicle.

D. A piece of a human Womb, most of whose interior surface has transformed into globular tumors.

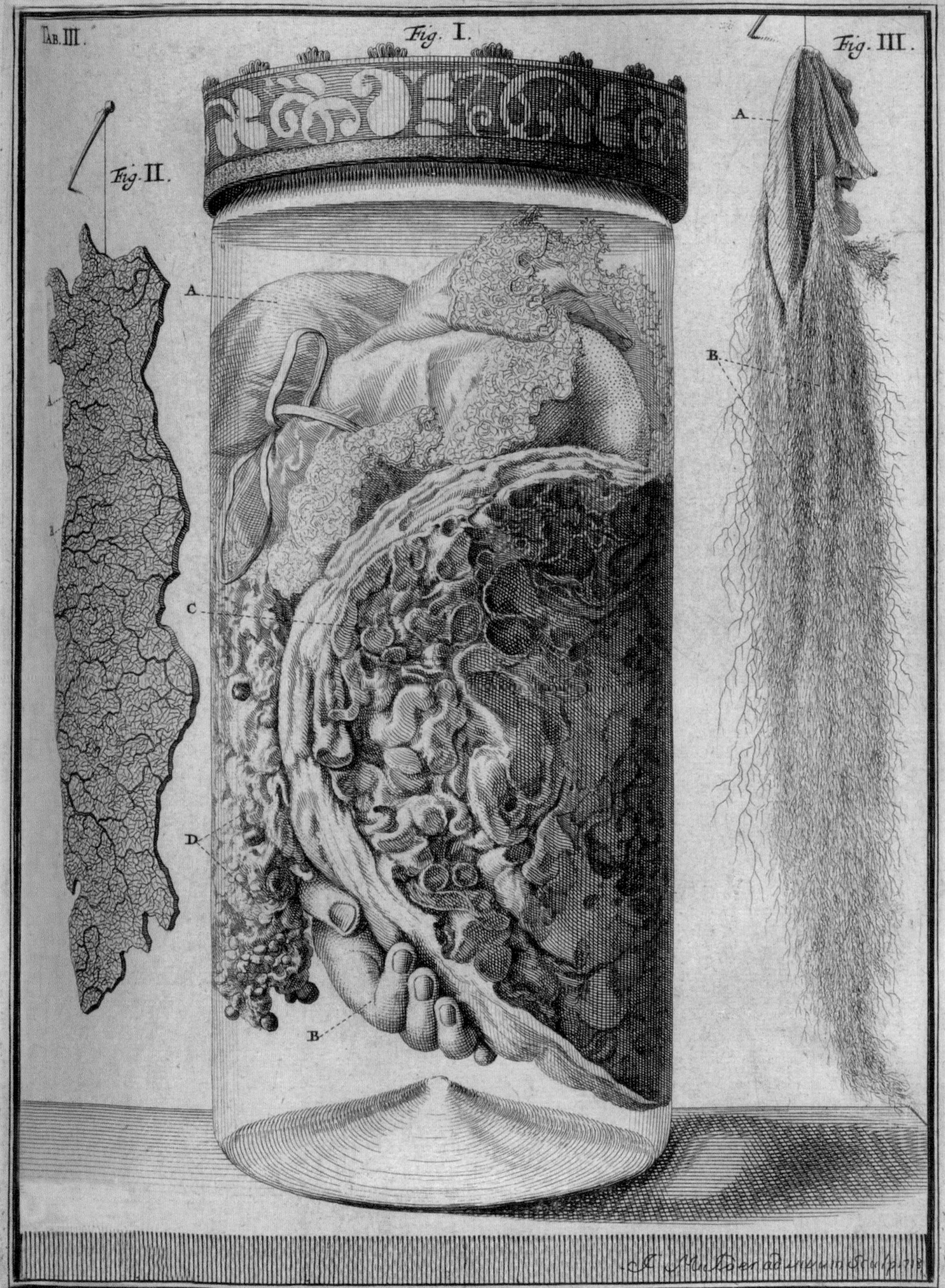

TAB. III.
Fig. I.
Fig. II.
Fig. III.
A.
B.
C.
D.

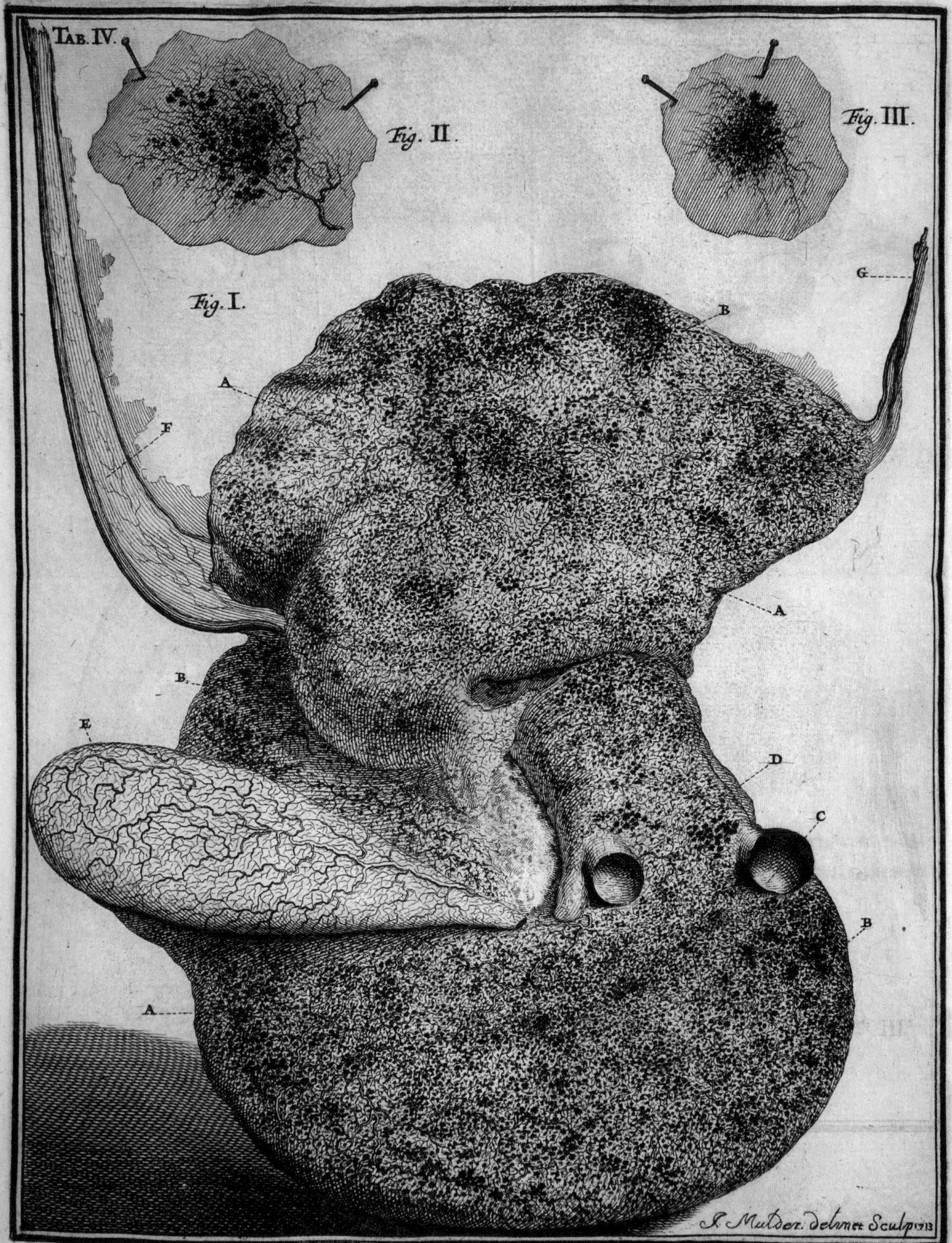

TAB. IV.
Fig. II.
Fig. III.
Fig. I.
G
B
A
F
A
B.
E
D
C
B
A
J. Mulder. delin et Sculp 1713

Fig. II. Shows a sample of human skin from the sole of the foot, whose blood vessels are so abundant that it is impossible to prick it with a lancet without hitting a vessel, but the illustrator was not able to represent this adequately.

A. Skin.

B. Vessels, primarily arterial, whose crawl clearly differs from those that make up the Spleen, the Kidneys, the Liver, the Placenta, &c.

Fig. III. Shows an unraveled human Testicle, whose vessels have been artificially emptied of their seminal material; having done this, the said vessels are now fine, like down feathers.

A. The white Membrane.[178]

B. The Vessels that make the rudiments of the seed.

Table IV [see image on previous page]

Fig. I. Shows the Liver of a youth.

A.A.A. The substance of the Liver,

B.B.B. The pulpy or juicy extremities, which are so very numerous that the engraver could hardly represent all of them here, & if he had, the entire Liver would be quite black with them.

C. The severed trunk of the Vena Cava.

D. The trunk of the Vena porta.

E. The gall Bladder, so abounding with Blood vessels that not even the point of a needle could be inserted between them.

F. The umbilical ligament, with its arteries filled.

G. The third ligament of the liver, which, in this place, contains the liver.

Fig. II. Shows a small piece of membrane surrounding the liver, below which can be seen the juicy & pulpy extremities, as seen under a microscope.

Fig. III. Shows a piece like the above, seen under such a lens that the pulpy extremities approach their natural size.

Table V [see image on facing page]

Fig. I & II. Two stones from a 68-year-old woman, removed by cutting, just as with a man.

Fig. III. Stones extracted in the same manner from a young woman, top view.

Fig. IV. The same stone, bottom view.

Fig. V. An Embryo of a Narwal, or Greenland Unicorn.

END.

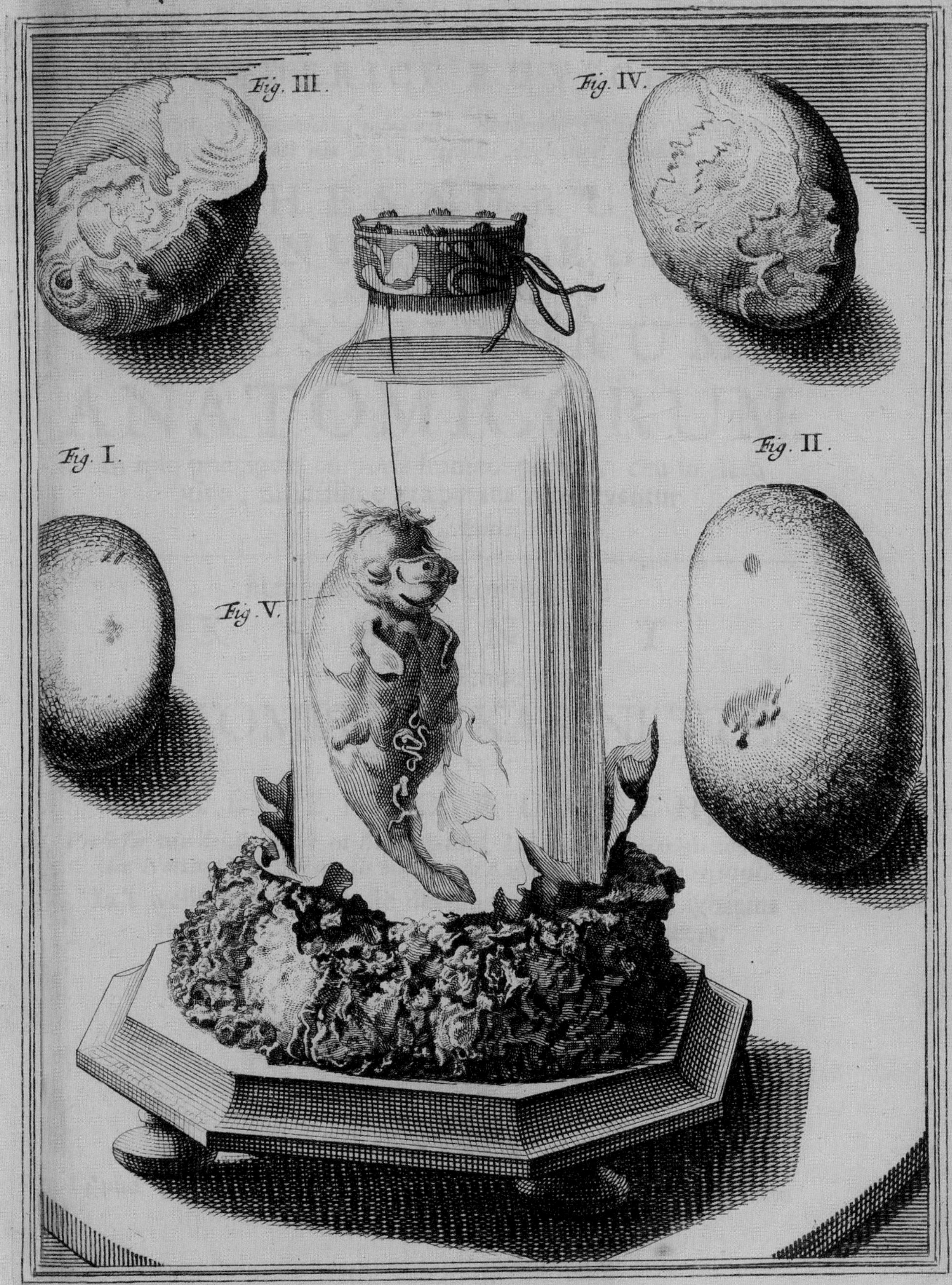
Fig. III.
Fig. IV.
Fig. I.
Fig. II.
Fig. V.

Frederik Ruysch

Professor of Anatomy & Botany, and Member of the Imperial Academy of Natural History

the Great & Royal Cabinet, that is
The Tenth Anatomical Cabinet
In which are preserved exceptional pieces of the
human body, so elegantly prepared as to seem alive.

With Engravings

Amsterdam, 1715.

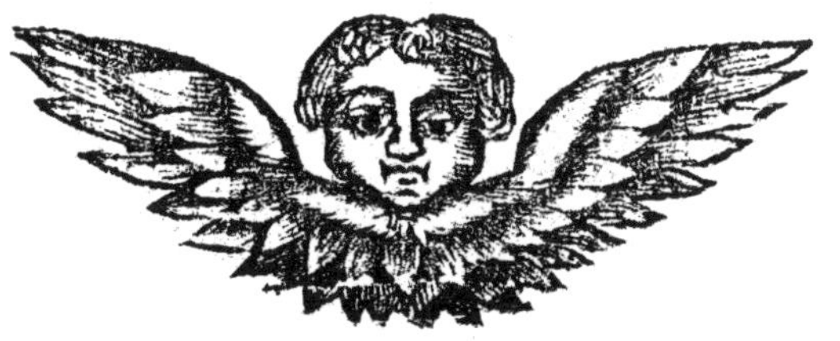

TO PATRONS OF MEDICINE, & PHILOSOPHY,
GREETINGS FROM FREDERIK RUYSCH.

When Andreas Vesalius[179] cast a dawning light upon the world of medicine, by committing to writing that incomparable work *On the Fabric of the Human Body*,[180] the best judges in the Art concurred that we now possessed a complete understanding of Anatomy: that, in fact, nothing of great importance remained for Posterity to add to his discoveries, and thereby win laurels of its own. They discovered, however, that the truth was far different from what they had supposed, just as soon as the most Meticulous Bartolomeo Eustachi[181] introduced his discoveries, and Gabriele Falloppio,[182] the foremost of Anatomists, publicized his own observations on Vesalius's work.

Of course Vesalius himself, that spotless soul, frankly acknowledged, in his many and most honorable responses to Falloppio's critiques, that he was indebted to the acute industry of his onetime pupil & present-day rival. But even then, in the wake of these Noble Innovators, the hope of discovering things most elegant & useful was not yet snatched away from the rest, as Giulio Casseri[183] soon showed with his outstanding advancement of the field, and Gaspare Aselli[184] as well.

Without doubt, William Harvey[185] provides another outstanding example that, even after so many brilliant discoveries, one person, possessed of inexhaustible diligence, may find things that all have missed, and that may also be of greater use for medicine than everything that all the others with their combined labor have discovered.

Indeed, the great splendor of this British star might have eclipsed many other lights, and yet he did not obstruct, so all the more could Pecquet,[186] Bartholin,[187] Lower,[188] Willis,[189] de Graaf,[190] Swammerdam,[191] Malpighi, Nuck,[192] & Du Verney[193] shine with his reflected radiance. What more is there to say? Driven onward by this subject, I have always believed that I should celebrate the work of others, in accordance with the Hippocratic Oath: so, indeed, with my industry, I also investigate their work, I further it, & when it falls to me to amend it, I venture to do as much as I can.

For this reason, even more than fifty years ago, in that season when de Bils[194] was, before all others, the Most Celebrated Anatomist, I already had the fortitude to scrutinize his opinions on my own, and to refute them with demonstrations of fact.[195] I ventured to describe the correct structure of the Lymphatic System, with the Lactic canals and the Bronchial Arteries; and then, in words and deeds, I publicly disclosed that I had discovered a technique, by which I could trace even the slightest crawl of the tips of those vessels, which I revealed and presented for examination.

From that time on, as the years raced by, I devoted my nights and days to the goal that, by employing all the force of my craft, I might obtain an understanding of that workmanship[196] God used to build us this machine.

And it pleased my Creator to crown this impertinent undertaking of mine with success, such that I at last conquered all obstacles with diverse skills and knowledge: that the pupil of my eye at last penetrated the deepest recesses of nature's inner sanctum: that after so much effort, and so many revisions, the true structure of the smallest parts at last lay open before me.[197]

At the very moment of my understanding, I arranged an exhibition of these shapes, at the first arising opportunity, for an expert audience, so they could examine my evidence: so it was that many witnesses, who had previously proven themselves candid judges on these matters, did not shy from using their intellect to challenge, to amend, & to acknowledge this new form.

So now to the present. With the growing weight of age denying me great expectations for further life, I had the idea to gather the finest examples of all my discoveries, and to organize them into a single collection, that would deserve the name The Ruysch Museum of Anatomy. In establishing this, however, I did not wish to follow my familiar path of merely writing; but I have put forward those very organs of the body, prepared by my art, whose contemplation provided the breakthrough in my understanding. Therefore I directed my enterprise from the outermost Hair all the way to the deepest secrets of the innermost viscera, nor did I omit anything of the middle parts between those extremes.

Thus, Fiber, Membrane, Vessel, Gland, Viscera, Cartilage, Bone, Nail, Hair—in short, every part, from head to toe, its true form exposed, might have a representative example.

I should perhaps add here that the little human cadaver has been prepared (by my own special technique), in such a way that its elegance & charm do not simply match life, but truly surpass it: for the color, plumpness, expression, & certainly the flexibility of limb which perish in corpses are all preserved here, & somehow even improved, so that the viewer is put at ease & becomes so beguiled that, unless convinced by their own eyes, no one would believe that this could have been achieved: and it greatly commends this practice that for more than fifty years the body has remained unchanged.

Lest any think this remarkable display will last only a few years, I have preserved all these items by means of my special embalming technique; so that they may endure a century, and still not decay. Moreover, I took care to mitigate the noisome appearance of parts taken from dismembered cadavers, with suitable clothing and other embellishments, so that the eye should in no way be troubled, and neither stricken with terror nor moved to nausea. Next, each part was assigned an individual number, that refers to a written catalogue, in which each number corresponds to an individual exposition that is clear, accurate, & easy to understand. And thus whoever explores this Museum through this catalogue will learn Ruyschian Anatomy, free from all squeamish activity, from the repugnance of filthy operations, and from foul stink.

Whoever possesses this Cabinet, then, may use it, either to amuse himself by contemplating the true construction of the human body, while at the same time improving his mind; or else, just as in a University, where a youth is educated in Philosophy & Medicine, he may be shown the genuine way the parts go together, and this will be done with neither words nor conjecture, but with an actual demonstration to look at.[198]

For this reason, Kings, Princes, or Administers of Republics or Academies, having acquired this Cabinet, may endow their Academies with a gift to endure for ages, an eternal & most useful instrument, as ornamental as it is instructive.

I, who have completed this, now offer it for sale myself, while I still am alive, so that, before I meet my ultimate fate, I may be sure that this record of my discoveries will remain intact: for that may not easily happen, once I am numbered among the many departed. This Cabinet is available for viewing at my house, by as many as provide proof that they are able & willing to purchase the complete collection.

If the buyer fear that these items will decay once he has purchased them, I promise in good faith that, after he collects his prize, I will disclose the technique of preservation that I use, & which differs from that which others employ: for any comparison of the things that I have preserved with the things that those others have proves that my technique is quite different. Likewise, I will show the method by which I prepared the liquid for preserving specimens in bottles, & the way it should be poured out, when needed.

To lessen the possibility of this need arising, however, I will refill the vessels, & close them properly, so that for many years they will require no further addition. Finally, since I know from experience that preparations from different parts of the human body must be preserved in different liquids, or else they will lose their luster, I will reveal those secrets, & others, to the Purchaser, under an oath of silence.

If, therefore, Princes, Academies, Associations, or private individuals are captivated by the sweetness of these things, they may purchase something that has been made by my art, a continuing labor of nearly sixty years; something that has no parallel anywhere else in the world; and that has never before been seen in any age.

EXPLANATIONS OF THE FIGURES TABLE I. [see image on facing page]

Fig. I. Shows a jar, in which a portion of the lower Jaw is preserved, with a piece of human Cheek.

A. Exterior covering of the Cheek.

B. Epithelia[199] of the Cheek.

C. Papillary nerves overlaid by the Epithelia, which are so exceedingly small that the engraver has not been able to outline them very well.

D. Decorative blue silk, placed beneath the jar.

E. Molars.

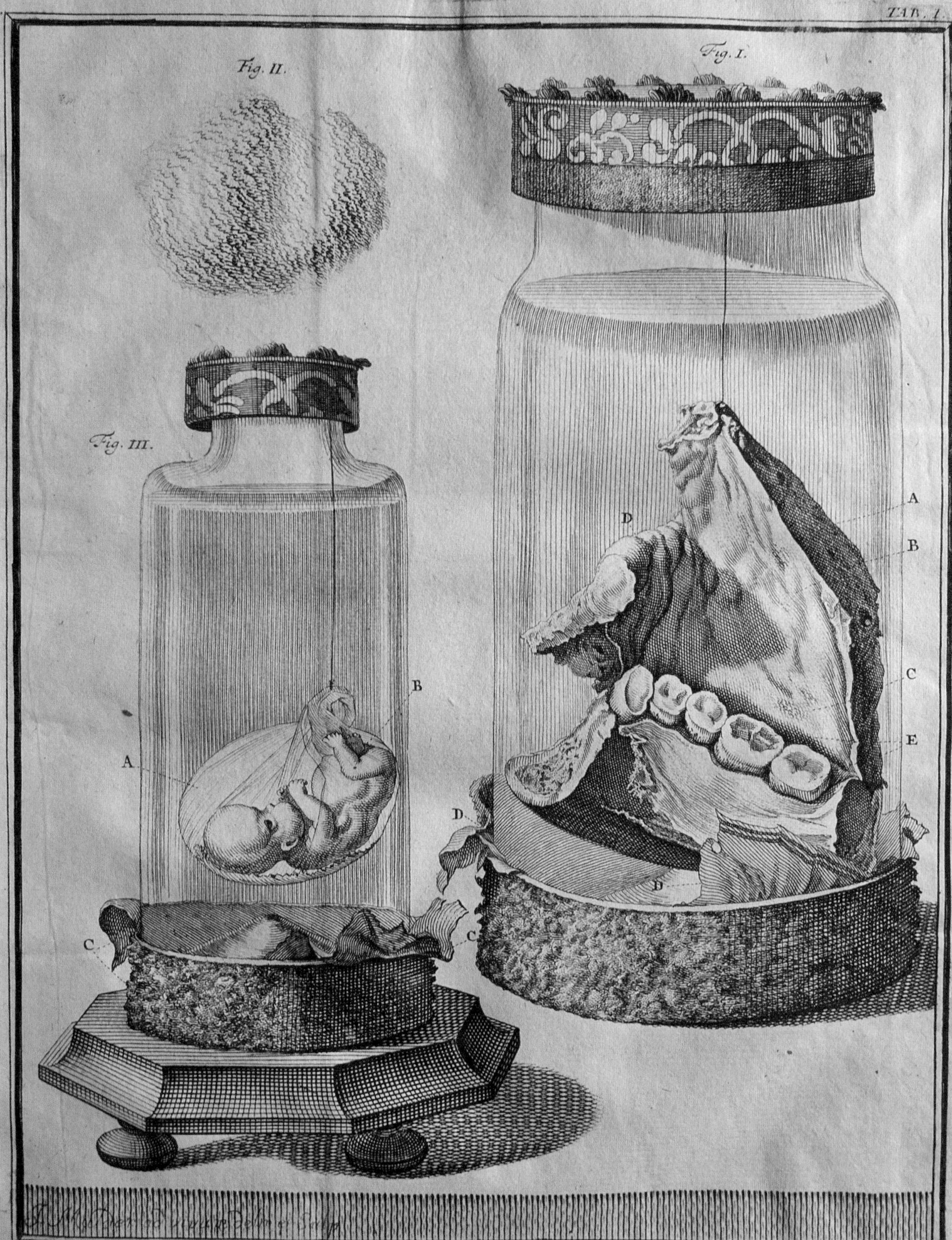
TAB. I
Fig. II.
Fig. I.
Fig. III.
A
B
C
D
E

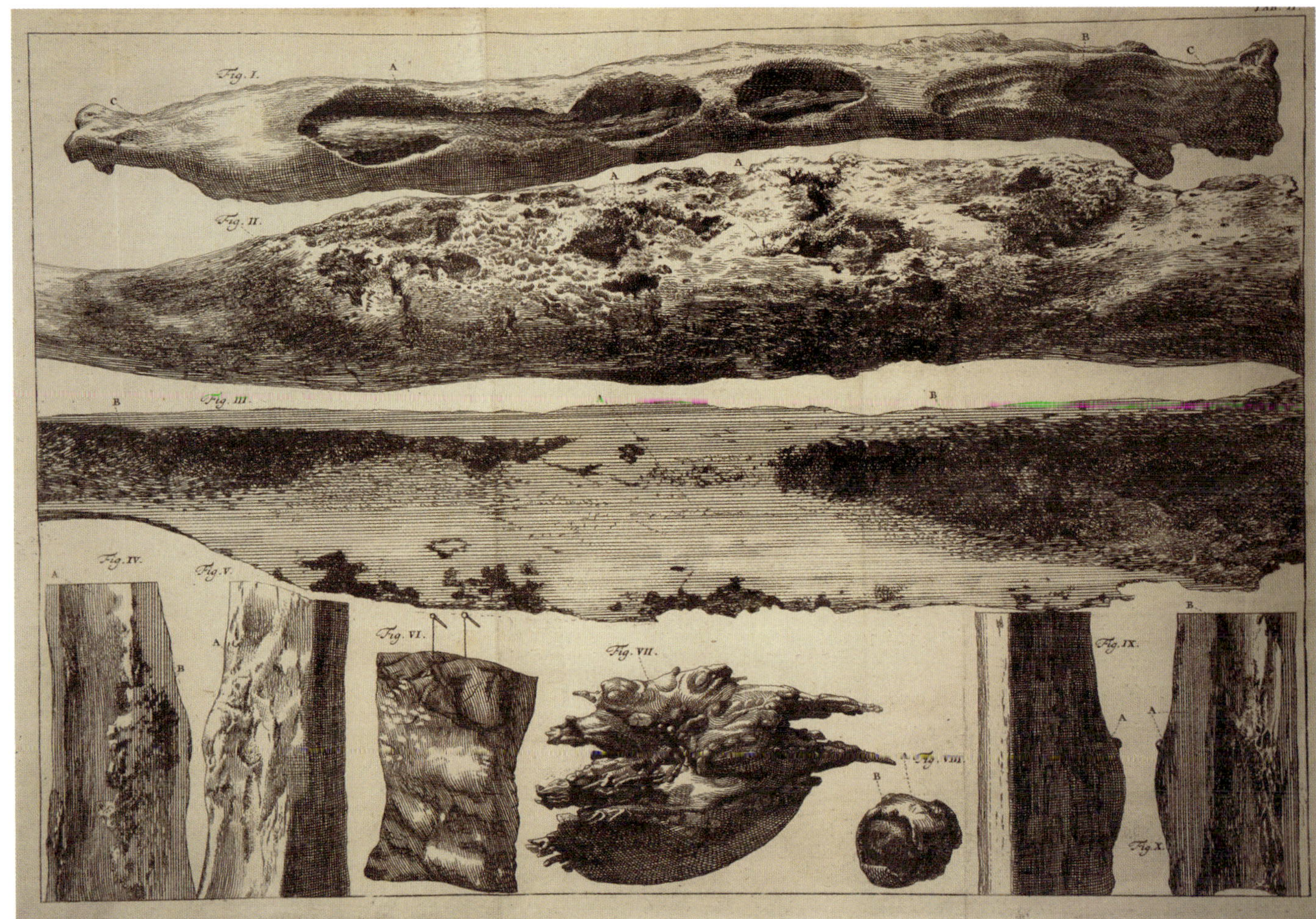

Fig. II. Shows a sample of papillary nerves from the Mouth, as seen under a microscope, whose lens is the size of a haddock's eye lens.

Fig. III. Displays a human fetus which is still contained in the Amnionic membrane, & even with its own fluid, in which it was hosted within the mother's body, while it lived.

A. Amnionic membrane, as delicate as spiderwebs, with its original fluid.

B. The baby, or Fetus.

C. A Coffer, & decorative blue silk, within which the crystal jar has been placed.

Table II. [see image above]

Fig. I. Displays the bone of a forearm, deformed by acrid humors.

A. A Fistula in the interior of the bone, which has separated all around from the exterior substance of the bone.

B. A pocket.

C. A deformed elbow.[200]

Fig. II. The outer face of a deformed Shin bone, divided into two parts by a saw.

A.A. Cavernous pockets.

Fig. III. The same bone, interior face.

A. A piece of bone, quite solidified, & and lacking its former cavity.

B.B. Osseo-spongiosus matter.

Fig. IV. A portion of Shin bone, covered with Tophus[201] bone & sawn in two.

A. The internal face.

B. An osseo-spongiosus bone spur, not particularly hard.

Fig. V. The same bone, exterior face.

A. An area of Tophus, full of hollows, & exceedingly hard.

Fig. VI. Skin, from the bone in the previous Figure, irregular & dreadful to behold.

Fig. VII. A piece of Elephant tooth, covered with fragments of tooth, perhaps caused by a gunshot.

Fig. VIII. A small piece of Elephant tooth, in which is found a copper ball, fired from a gun.

A. The piece of Tooth.

B. The copper Ball.

Fig. IX. A piece of Shin bone covered with tophus, & divided lengthwise.

A. The external face near a mound of tophus.

Fig. X. The same bone, interior face.

A. Bony substance, exceedingly thick & hard.

B. A bone spur with crevices.

Table III. [see image on page 225]

Fig. I. An embryo hanging from its umbilical cord.

Fig. II. A piece of shoulder[202] Bone, from a little boy.

A. The upper epiphysis, consisting partly of cartilage and partly of bone.

B. Bony particles.

C. The interior substance, which is spongiform bone, imbued with liquid marrow, & still covered with membrane so thin as to match a spiderweb. This boasts many filled arterioles, & therefore flushes red.

Fig. III. A piece of Skin from a Black African.[203]

A. Exceedingly thick skin.

B. A body of Papillary nerves.

Fig. IV. A piece of Thigh bone from a little boy, in whose upper Epiphysis the cartilage and various bony particles can clearly be seen.

Fig. V. The surrounding Membrane of a human Liver, through which crawl countless Arterioles.

Fig. VI. A portion of the Forehead of a boy, throughout whose upper Eyelid are scattered Arterioles, issuing from the inner corner of the eye.

END.

Fig. I.

Fig. II.

A

C

B

Fig. III

B.

A.

Fig. IV.

Fig. V.

Fig. VI.

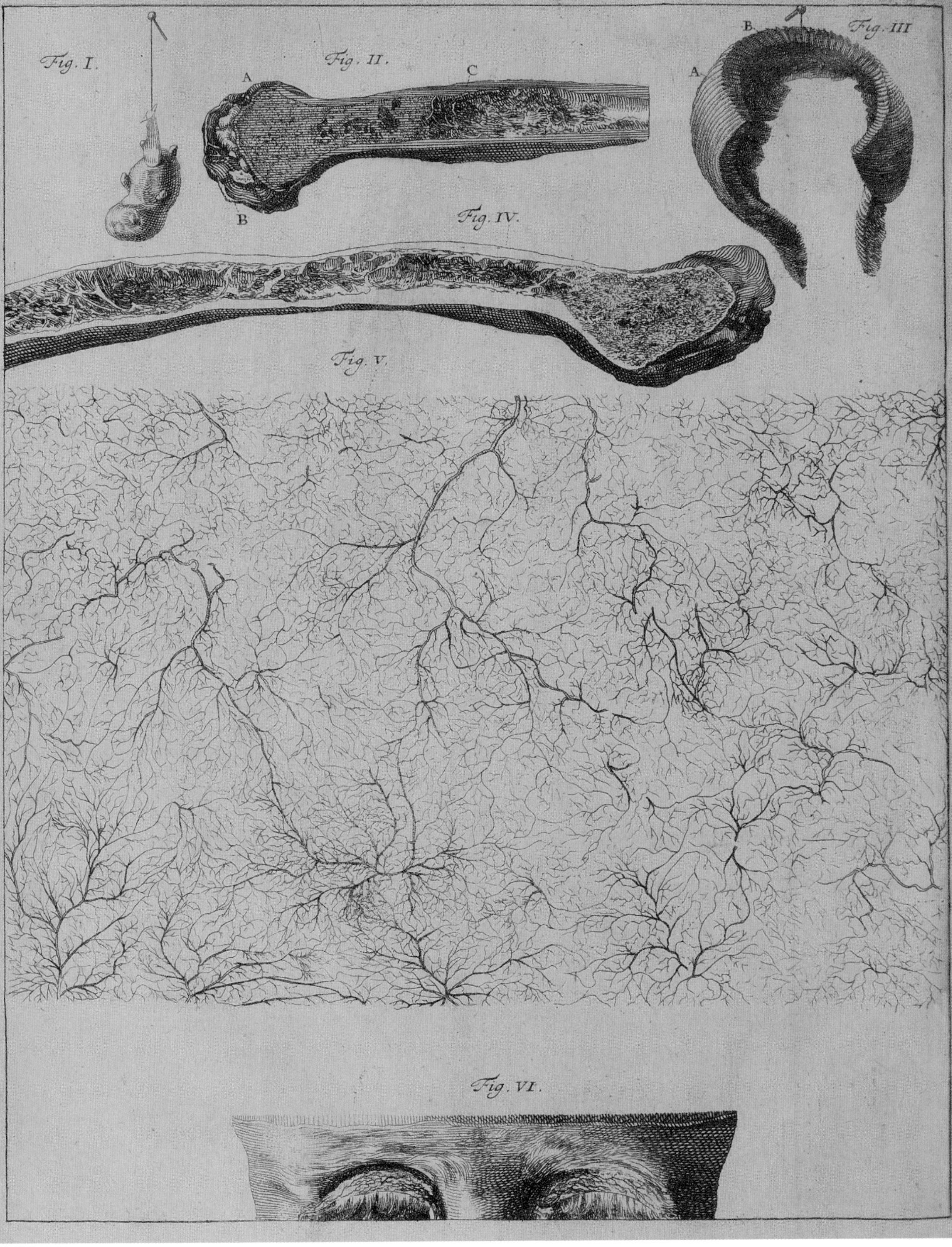

Notes

Timeline sources: http://galileo.rice.edu/Catalog/NewFiles/ruysch.html; https://embryo.asu.edu/pages/frederik-ruysch-1638-1731.

1 L. P. Sue, "Discours historique sur la vie et les ouvrages des citoyens Sue," in *Séance publique de l'Académie de Chirurgie* (Paris: chez Croullebois, 1793).

2 A. Karr, *Les guêpes*, 5th series (Paris, August 1848), 3.

3 P. Vallery-Radot, *Chirurgiens d'autrefois, la famille d'Eugène Sue* (Paris: Ricou-Ocia, 1944), 50–54.

4 L.-V. Thiéry, *Guide des amateurs et des étrangers voyageurs à Paris* (Paris, 1787), 404.

5 *Le Moniteur*, April 15, 1819.

6 Archives Nationales, AJ52 13.

7 Ibid., AJ52 450.

8 J.-J. Sue the Younger, *Extrait du discours prononcé par M. le professeur Sue, le jour de l'installation de son muséum dans la nouvelle salle destinée au cours d'anatomie pittoresque* (Paris, [1825]).

9 Alexandre Dumas, *Les morts vont vite* (Paris, 1861), vol. II, 20.

10 Archives Nationales, AJ52 450.

11 Ibid., AJ52 450.

12 Ibid., AJ52 809.

13 Ibid., AJ52 443.

14 Ibid., AJ52 450.

15 Michel Lemire, *Artistes et mortels* (Paris: Chabaud, 1990), 232.

16 See Luuc Kooijmans, *De doodskunstenaar. De anatomische lessen van Frederik Ruysch* (Amsterdam: Uitgeverij Bert Bakker 2004); and Jozien J. Driessen van het Reve, "Frederik Ruysch: revolutie op kousenvoeten," in O. P. Bleker and Jozien J. Driessen, *Geloof alleen je eigen ogen: een actuele kijk op de anatomische preparaten van Frederik Ruysch (1638–1731)* (Hilversum: Uitgeverij Lias, 2017).

17 Kooijmans, *De doodskunstenaar;* Driessen, "Frederik Ruysch: revolutie op kousenvoeten."

18 Kooijmans, *De doodskunstenaar*; Driessen, "Frederik Ruysch: revolutie op kousenvoeten."

19 See Bert Theunissen and Robert Paul Willem Visser, *De wetten van het leven:*

historische grondslagen van de biologie 1750–1950 (Baarn: Ambo, 1996); Robert J. Richards, *The Romantic Conception of Life Science and Philosophy in the Age of Goethe* (Chicago: University of Chicago Press, 2004).

20 Timothy Lenoir, *The Strategy of Life: Teleology and Mechanics in Nineteenth-Century German Biology* (Chicago: University of Chicago Press, 1998).

21 For this, however, it was essential to see embryological development as epigenesist: a process in which more complex structures would only gradually develop from simpler ones until development had completed. This view is normal to us now, but in the life and times of (for instance) Ruysch a competing view on embryology was more in vogue, namely preformation: the view that the embryo—a mini-human—was already fully formed at its start; it only needed to grow. Ruysch's friend Jan Swammerdam, famous for his studies and preparations of developing insects, was one of its advocates. Preformation fitted into the mechanical worldview of the seventeenth century, but, as with the newer concept of vitalism, it had gradually been replaced by epigenesis in the second half of the eighteenth century.

22 See also Laurens de Rooy, "A Cabinet Departs," in Laurens de Rooy and Hans van den Bogaard, *Forces of Form* (Amsterdam: Vossiuspers UvA, 2009).

23 Willem Vrolik, *Het leven en maaksel der dieren* (Amsterdam, 1853).

24 See Hieke Huistra, *Preparations on the Move: The Leiden Anatomical Collections in the Nineteenth Century* (Leiden: Proefschrift Universiteit, 2013).

25 See Anne Harrington, *Reenchanted Science: Holism in German Culture from Wilhelm II to Hitler* (Princeton: Princeton University Press, 2009).

26 See David Baneke, *Synthetisch denken: natuurwetenschappers over hun rol in een moderne maatschappij,* 1900–1940 (Hilversum: Verloren, 2008).

27 "[V]ele mogen thans de steenenbrengers zijn, weinig zijn de bouwmeesteren." See Laurens de Rooy, *Lodewijk Bolk en de Bloei van de Nederlandse anatomie* 1860–1940 (Amsterdam: Amsterdam University Press, 2011).

28 Lodewijk Bolk, *Hersenen en Cultuur* (Amsterdam: [Verm Herdr], 1918).

29 "Elke vorm waarin het leven zich hult, draagt den stempel der individualiteit": in Bolk, *Hersenen en Cultuur.*

30 "[W]ij zijn gewoon het leven na te sporen door vergrootglazen en daardoor het anders onzichtbaar stoffelijke binnen onzen gezichtskring te brengen, hoe geheel anders … zou onze opvatting van het leven zijn, indien het ons gegeven ware, dit eens te bestudeeren met verkleinglazen, waardoor wij het voor het ongewapende oog onoverzichtbare binnen onzen gezichtskring konden brengen, om dan in plaats van zooals thans de stoffelijke verbindingen, den samenhang der verschijnselen meer tot studiedoel te nemen." In Bolk,

Hersenen en Cultuur.

31 Steiner, an early twentieth-century esotericist, hypothesized that there existed a spiritual world which could be known and understood by humans.

32 This is my own translation of the well-known Latin passage.

33 "The Chemistry of Isaac Newton," http://webapp1.dlib.indiana.edu/newton/mss/Dipl/ALCH00017. Newton even did his own translation of the *Emerald Tablet*, which was found much later in his alchemical notes.

34 Cited in Philip Ball, *Curiosity: How Science Became Interested in Everything* (Chicago: University of Chicago Press, 2012), 57.

35 Michael E. Moran, *Urolithiasis* (New York: Springer-Verlag, 2016).

36 David Mazierski, "The Cabinet of Frederick Ruysch and the Kunstkamera of Peter the Great: Past and Present," *Journal of Biocommunication* 38 (2012): E31–E40.

37 Anthony Anemone, "The Monsters of Peter the Great," *Slavic and East European Journa*l 44, no. 4 (2000): 583–602.

38 "Acquisition of Collections in Europe: Frederik Ruysch, Albert Seba, and Joseph-Guichard Duverney," http://www.kunstkamera.ru/en/museum/kunst_hist/5/5_2r.

39 Frederik Ruysch, *Dilucidatio valvularum in vasis lymphaticis et lacteis cum figuris aeneis, accesserunt quaedam observationes anatomicae rariores* (The Hague: Herman Gael, 1665).

40 See the excerpt of Campanella's *On Sense and Magic* in Brian P. Copenhaver, *The Book of Magic: From Antiquity to the Enlightenment* (London: Penguin, 2017).

41 Frederik Ruysch, B. Cole, and Godfrey Lowe, *The Celebrated Dr. Frederic Ruysch's Practical Observations in Surgery and Midwifry* (London: Printed for T. Osborne, 1751).

42 Julie V. Hansen, "Resurrecting Death: Anatomical Art in the Cabinet of Dr. Frederik Ruysch," *Art Bulletin* (1996), 663–679.

43 Lucas Boer, Anna B. Radziun, and Roelof-Jan Oostra, "Frederik Ruysch (1638–1731): Historical Perspective and Contemporary Analysis of His Teratological Legacy," *American Journal of Medical Genetics, Part A* 173A, no. 1 (2017): 16–41.

44 Max Weber, *Essays in Sociology*, ed. and trans. H. H. Gerth and C. Wright Mills (New York: Oxford University Press, 1946), 129–156.

45 Detractors accused Ruysch of nothing more questionable than using his aesthetic embellishments to conceal the shortcomings of his preservation technique.

46 That two of his daughters were accomplished painters argues that the arts were respected in the Ruysch household.

47 I have abridged parts, however, and have marked those places with footnotes.

48 See the preface to the Second Anatomical Cabinet.

49 "Anatomica." The word still retained the force of its original Greek meaning: "cut apart."

50 At this time, nerves were believed to be hollow conduits, like blood vessels.

51 "Taedio" refers to the feeling of repugnance caused by the monotonous repetition of work that is inherently irksome or disgusting. Hence, its meanings can range anywhere from boring to depressing to revolting. Here, the balance might tilt toward boredom, since Ruysch's topic is the voraciousness of his intellect. But he can be quite frank about the filth, stench, and emotional distress that accompanies the dissection of corpses. (See the preface to the Tenth Cabinet.)

52 I have supplied that word. Beyond "parts" and, rarely, "objects," Ruysch himself does not seem to have a generic term for the items in his collection. The structure of the Latin language allows for this vagueness. Verbs do not require explicit subjects and adjectives can stand alone, so it is not strictly necessary for Latin writers to name what they are discussing. Where I may have used "specimens" or "items" Ruysch generally writes "these" or "those" or relies on an implied "things."

53 That becomes the subject of Ruysch's *Thesaurus animalium primus.*

54 "Deo Ter Opt. Max." Ruysch uses this convention from classical Latin. If we want to be more faithful to the text (and somewhat more exotic), it could be rendered as "the Thrice-Best and Greatest God."

55 "Ligno Indico," another name for guaiacum, or lignum-vitae, one of the hardest of woods, which is indigenous to the West Indies and South America. The sap was considered a cure for a variety of ailments, including gout, kidney stones, and syphilis.

56 According to the numbers Ruysch gives above, he had 14 cabinets. In volumes two and six, he puts the number at 16. Ultimately, he would publish descriptions of 10. After selling his collection to tsar Peter the Great in 1718, he would publish two more volumes on anatomical cabinets he curated later.

57 Lithotomy was one of the most common surgical procedures of the time. The physician would locate the stone by means of a long metal sound, inserted up the urethra and into the bladder. While there were several methods for removing the stone, they all involved opening the bladder from between the

legs and the peritoneum.

58 The uterus had descended through the vagina and was visible between the labia.

59 A funnel-shaped cavity within the kidney that conveys urine to the ureter, the tube leading to the bladder.

60 "Homo natus de muliere, brevi vivens tempore, repletur multis miseriis." Job 14:1.

61 "tria Sceleta foetuum abortivorum quator circiter mensium." "Abortivus" (prematurely born) covers premature births, miscarriages, and intentional abortions.

62 "Nec parcit imbellis juventae poplitibus." Horace, *Odes*, Book III, Poem 2.

63 "Cur ea deligere velim, quae sunt in Mundo?" Unidentified.

64 None of the remaining items on this shelf are illustrated.

65 "Asperis Arteriis": this includes the trachea and bronchial tubes.

66 See below.

67 From antiquity, Democritus and Heraclitus were commonly contrasted as the laughing philosopher and the weeping philosopher, although the distinction is not entirely supported by their work.

68 "Cum tantum in vita restet transire malorum, / Liber ab his letho, laetus sine voce triumpho." Unidentified.

69 Virgil, *Aeneid*, Book VI, 428–429: "dulcis vitae exsortes, et ab ubere raptos / Abstulit atra dies, et funere mersit acerbo."

70 "I will praise thee; for I am fearfully and wonderfully made: marvelous are thy works; and that my soul knoweth right well. / My substance was not hid from thee, when I was made in secret, and curiously wrought in the lowest parts of the earth."

71 "duas Rupes": *rupes* ordinarily means "rocky cliffs," but Ruysch seems to be using it to mean something like "landscape" or "miniature scene."

72 "Capitis ossei,"

73 The aorta.

74 Today also called the internal thoracic artery.

75 "Omnia sunt Humanum tenui pendentia filo." Ovid, *Epistolae ex Ponto,* IV.3.35.

76 There are twelve pair of cranial nerves. As far as I can tell, until at least the mid-eighteenth century, only eleven had been identified.

77 The vestibulocochlear nerve.

78 A thick membrane that encloses the brain and spinal cord. The four folds Ruysch references separate various sections of the brain.

79 The sella turcicia (Turkish saddle), a hollow in the skull that holds the pituitary gland.

80 Perhaps the glossopharyngeal nerve. The only other cranial nerve that Ruysch fails to mention is the facial nerve.

81 On either side of the head, each of the two vagus nerves exits the cranium through a hole called the jugular foramen, next to which is a smaller hole called the foramen spinosum.

82 A fold in the peritoneum that holds the small intestine in place.

83 See *Epistola anatomica, problematica quarta,* in Frederik Ruysch, *Opera omnia* (Amsterdam, 1721), vol. 1, p. 490.

84 From Stephen Blancard, *The Physical Dictionary*, 4th ed. (London, 1702), 248: "Polypus, a Swelling in the hollow of the Nostrils, and is twofold, either like a Tent, and goes by the general name of *Sarcoma*, or such a one that has a great many distinct Branches or Feet which extend either to the outside of the Nose, or in the inside of the mouth: Their Colour is White, often times reddish, and sometimes Black and Livid. Excrescencies of this nature happen not only in the Nostrils, but sometimes in the Heart and Cavities of the thicker Membrane of the Brain."

85 Ruysch's word to describe the shape of the tips of the blood vessels is *reptatus*, which means "a crawling or creeping." As far as I can tell, it is idiosyncratic to him. The descriptiveness of the term reflects his obsession with capturing the shapes of all the various vessels.

86 Descriptions of Tables I and II are in the preceding text.

87 Papillary nerves get the name from tiny fingertip- or budlike structures, which "are the organs of touch, being the termination of the cutaneous nerves." (G[eorge] Motherby, *A New Medical Dictionary, or General Repository of Physic* [London: J. Johnson, 1785], 559; cited as Motherby in subsequent notes.) Since the Latin word for nipple is *papilla,* Ruysch has to distinguish between the nerve papillae of the skin and the papillae of the breast. The heavy repetition of inflected forms of *papilla* throughout the original Latin on this page certainly creates a rhetorical effect, but to what end is far from clear.

88 "True skin" to denote in particular the layer below the epidermis, rather than the skin in the general sense.

89 A thin layer of membrane that envelops the brain.

90 One of Ruysch's critics cites this technique as a cheat to make corpses

appear more pristinely preserved. See the preface to the Ninth Cabinet.

91 "Papillae," the same word for "nipples," which likely explains why Ruysch chose to include this specimen here.

92 For discussion of glands, see note 46.

93 The layer of skin below the epidermis.

94 Since Ruysch makes slighting references to glands throughout the *Thesaurus anatomicus*, some background might be helpful. Until roughly the mid-eighteenth century, physicians still subscribed to the doctrine of humorism, which was most famously articulated in the second century by Galen, the celebrated surgeon and medical writer. By Ruysch's time the theory had been extensively modified, but the central tenet held that each part of the body had its particular nutritional requirements, which were supplied by certain vital fluids. These fluids were made up of differing proportions of individual "juices," or "humors," which were separated out (or "secreted") at the appropriate location.

Many medical authorities argued that this secretion was performed exclusively by "glands," located at various points, including the tips of the vessels. Ruysch's first significant contribution to medicine was his demonstration that structures in the lymph system that had been taken for glands were in fact valves. From here, he would go on to show almost a personal animus toward the idea of filtering glands. Instead, he advanced the idea that it was the tips of the vessels themselves that filtered out the various humors.

Aside from the aesthetic and technological pride Ruysch clearly felt for the wax injection technique he eventually perfected, he believed that it represented a major advance in anatomy because it revealed that the vessels simply tapered away and were not capped off by glands.

95 The text for this and the next item are more complete descriptions of the items illustrated in Tables III and IV of this cabinet.

96 These are technical terms referring to fundamental varieties of tissue. Since different schools of thought identified anywhere from three to a dozen or more basic tissues, the terminology can be inconsistent and difficult to nail down.

97 In the original text, all references in this section incorrectly indicate Table III.

98 "periostio."

99 "epiphyses."

100 The sclera, or white of the eye.

101 "Ruyschiana," i.e., the tunica ruyschiana, a layer lining the choroid coat.

102 Also known as the choroidea, it is the vascular layer of the eye between the sclera and the retina.

103 "ex homine": although this could indicate either a man or a woman, Ruysch tends to specify when his specimens come from women.

104 According to Motherby (355), these drain excess tears from the eye and out to the sinuses.

105 Motherby, 255: "It is the circular line on the globe of the eye, where the sclerotis, choroides, retina, cornea, processus ciliares, and iris terminate, forming a whitish ring …"

106 Motherby, 258: "It is a range of black fibers, circularly disposed, having their rise in the inner part of the choroides …"

107 "circulus Iridis minor," one of two arteries that together form a ring around the iris.

108 The eye lens.

109 Mislabeled as "E"

110 The original provided a reference to a passage that I have not translated.

111 Although there are nine figures in this table, there are only eight descriptions.

112 A vein that carries blood away from the heart, in this case to the liver.

113 A vein that carries blood back to the heart.

114 "transparentia," perhaps a typo for "transportentia," in which case the sentence would read, "and these are for transporting and preparing juice for nourishing the fetus, while lodging in the womb."

115 This difference in structure lays the groundwork for Ruysch's hypothesis that the secretion of the humors is produced by variations in shape at the tip of the vessels.

116 An external membrane of the fetus.

117 Those illusory "specks," Ruysch implies, are what have been taken for humor-secreting glands.

118 "Petrus Francius." Unidentified.

119 "Quid sumus, aut quidnam nostri post funera restat? / Adspice: Nil aliud, nuda nisi ossa, vides. / Calculus, et vastus lapidum, quem cernis acervus, / Quam sit vita hominum plena dolore, docet. / Exhibet hoc tabulis, hoc veer-

bis explicat, Urbis / Ruuschius Amstelliae laus, medicumque decus."

120 Many understand this skeleton to be playing a violin. While the illustration is highly suggestive, the text itself is ambiguous: Ruysch could well be playing on "expressing" musical emotion and "pressing out" the decay from the bone, but he could also be alluding to the cancer of Original Sin, and the piece of decayed bone "expelled" from its femur could "express" an analogy to fallen humankind. Both readings feel strained.

121 See the second cabinet, Table II, Fig. 2, as well as the text for the second shelf, item no. VI, above.

122 In general, bezoars are stonelike lumps formed in the stomach out of undigestible material, usually hair. From Motherby, 167: "[An oriental bezoar] is supposed to be produced in the cavity at the bottom of the fourth stomach of a species of goat in Persia called parau. It is only found in the old ones, and only in those which feed on particular mountains. ... The genuine ones are about the size of a kidney-bean, generally of an oblong rounded figure, but often of other forms, varying according to the shape of the body over which they are as a crust; for all the *bezoars,* whether animal or fossil, are formed upon a nucleus of one kind or another, as a bit of straw, a stone, or other matter. They have an even smooth surface and are of a shining olive colour, or of a dark greenish one." Powdered oriental bezoar was considered an antidote to poison.

123 See Ruysch, *Opera omnia*, vol. 1, pp. 479ff.

124 Since capital *I* and *J* were not differentiated yet, the list goes from I to K.

125 "meconium."

126 "Prima, quae vitam dedit, hora carpsit." Seneca, *Heracles Furens*, line 875.

127 "Quasi solstitialis herba paulisper fui: / repente exortus sum, repente occidi." Plautus, *Pseudolus*, 1.1.38–39.

128 "Tunicam Vaginalem": a pouch of membrane that surrounds the testicle.

129 A thin layer of muscle that can partially raise the testicle.

130 "O, quam dura premit miseros conditio vitae!" From Maximianus, a sixth-century Italian considered one of the last Latin poets.

131 This refers, not to the nation, but to one of the five races recognized at the time—American, Caucasian, Ethiopian, Malayan, and Mongolian.

132 Descriptions of Tables I and II are in the preceding text.

133 The corpora cavernosa, the two larger regions of erectile tissue in the penis.

134 "Corpus spongiosum" (the spongeous body), the smaller region of erectile tissue that surrounds the urethra.

135 Motherby: "The true skin on its whole surface is covered with two lamellae [layers], one is the *rete mucosum* [also called *corpus reticularis*], the other is the cuticula. The *rete mucosum* is the principal seat of colour in man."

136 From Motherby, 621: "It is a kind of tumor, thus named from its poultice-like contents."

137 The omitted text is a lengthy, technical note that refers to an untranslated part of the text.

138 "Crista Galli," a bone in the nose.

139 "Perichondrium."

140 "Plica Polonica." From Motherby, 589: This "consists of several blood vessels running from the head into some of the *hairs*, which cleave together, and hang from the head in broad flat pieces. ... They are painful to the wearer, and odious to every spectator."

141 The outer membrane of the fetus.

142 One of the membranes of the afterbirth.

143 Ruysch takes pains to explain this, because folk wisdom held that these hardened clumps of placenta were the remains of malevolent, literally misbegotten creatures. See Ruysch's *Observationes* XXVIII, XXIX, LVIII.

144 Two protuberances on the femur to which the muscle is attached.

145 A region of the dura mater.

146 Motherby, 382: "the membranous part which is found in new-born infants at the coronal and sagittal commissures, and which, in length of time, hardens into bone." That is, the soft spot of the head.

147 Part of the *processus falciformis*.

148 The pointy projections from the back of the vertebrae.

149 Small bones of the foot.

150 A disease also known as tuberculous dactylitis.

151 I have translated most of this very long foreword because it is revealing of Ruysch's personality and professional concerns and because it also gives a vivid account of the arena in which science was contested in his era. With that in mind, I cut many medical details, which doubtless from Ruysch's point of view constituted the heart of his argument, but which might try the patience a nonexpert audience.

152 Raymond Vieussens (d. 1715), a French anatomist who did pioneering work on the circulatory system.

153 "Novi systemmatis Vasorum sanguineorum in corpore humano." The book, published in 1705, is usually known as *Novum vasorum corporis humani systema* (A new system of human vessels).

154 Throughout Ruysch is using forms of *invenio*, which originally meant "I find." Only in the Renaissance did the word start to acquire the meaning "invent," in the modern sense of creating something entirely new. This shift in meaning reflected an ongoing intellectual and theological debate over whether it was possible, through art or technology, for humankind to originate anything or whether, to the contrary, artists and scientists were simply finding, or at most recombining, elements that had existed all along. How we choose to understand *invenio* therefore implies an assessment of Ruysch's self-perception: when he formulated his preservation technique, did he view himself as a diligent *discoverer* of the secrets of nature or a genius *inventor* who had created something utterly without precedent?

155 "Mortuus, Arte Tua, Ruyschi, vivit, docet, infans, / Elinguis loquitur, mors timet ipsa sibi."

156 Specifically, Ruysch means that the capillaries empty after death and become impossible to see.

157 Jacob Hovius (1675–1740).

158 Soft and full of tiny vessels, the brain had proven a major challenge to preparators, so from a purely technological perspective Ruysch was justifiably proud of his achievement. But, as this and earlier Cabinets make clear, Ruysch was also deeply invested in dispelling the belief that glands were the sole agents of secretion.

159 There are in fact seven figures.

160 Perhaps *Archilochus colubris*, the ruby-throated hummingbird.

161 The Table contains eight figures, but the last two are not mentioned in the text.

162 Motherby, 589: "Le Dran explains it to be a folding of the carotid artery in the brain."

163 One of the two layers of the dura mater.

164 Cells of the outermost lining.

165 "intestini Recti."

166 Gives the reference: "folio. Hort. Amst. part. I," which I have not been

able to identify.

167 See the artificial polyp, cabinet I, shelf I, item III.

168 "Diverticulum."

169 The text reads "immutari" (to be altered). I have taken this as a misprint for "immurari" (to be enclosed in walls).

170 1659–1734.

171 Playing on an aphorism that it was no less honorable to earn money than to inherit it.

172 "Sit minor virtus, quam quaerere, parta tueri: / Ars tua, Magne virum, praestat utrumque simul, / Immortale tuae nomen virtutis & artis / ILLIVs DoCtae faMa sVperstes erIt."

173 "Corporis humani fabricae quicunque stupendae / Vsurpare oculis interiora cupit, /Thesaurus adeat, RUYSCHII quos ingeniumque / Dexteritasque manus non imitanda struunt."

174 "Henr. Christianus Krugerus, Lunenburgensis."

175 Johann Jakob Rau (1668–1719), a colleague and rival.

176 "Cuticula," i.e., the epidermis.

177 "ceratholithophyti," i.e., *Ceratophyllum*, a common genus of water plant. *Ceratophyllum* literally means *hornleaf*, so Ruysch's formulation (literally *hornstoneleaf)* is possibly a play on words, but it is more likely an obsolete or idiosyncratic name for one of the many species of this plant.

178 *Tunica albuginea,* a fibrous covering inside the testis.

179 Andreas Vesalius (1514–1564), a Flemish physician considered the founder of modern anatomy. Because he is still best known by the Latinized form of his name, I have used that rather than the Dutch form, Andries van Wesel. All other names, however, I have de-Latinized.

180 "de Fabrica corporis humani" (more usually *De humani corporis fabrica*), the foundational text of modern anatomy, published in 1543. The Latin word *fabrica* means "craft" or "workshop," and, when Ruysch uses it elsewhere, that's how I translate it. However, since Vesalius's book is commonly known as *De fabrica*, I thought any translation other than "fabric" would be unnecessarily confusing.

181 Bartolomeo Eustachi (c. 1500–1574), a contemporary of Vesalius recognized as one of the founders of modern anatomy.

182 Gabriele Falloppio (1523–1562), another influential early anatomist.

183 Giulio Cesare Casseri (1552–1616), anatomist and author of *Tabulae anatomicae.*

184 Gaspare Aselli (1581–1625) discovered the lacteal vessels of the lymphatic system.

185 William Harvey (1578–1657), physician to James I of England. With the publication of *De motu cordis* (1628), Harvey became the first to accurately describe that the blood vessels formed a closed circulatory system, set in motion by the pumping action of the heart. This was utterly at odds with the traditional understanding, inherited from the Roman physician Galen, that blood ebbed and flowed through the vessels and ultimately remained in the extremities, where it was converted into tissue. While Harvey's theory initially faced stiff resistance, it was ultimately hailed as a major advancement in the understanding of the human body.

186 Jean Pecquet (1622–1674) researched the lymphatic system in animals. Most of the other names in this list are mentioned in Motherby as also having done important research on the lymphatic system, which was of particular interest to Ruysch, too. This illustrates how, following Harvey's work, completing the understanding the vascular system and the secretion of the body's various vital fluids became a major project for anatomists.

187 Thomas Bartholin (1616–1680) did pioneering work on the human lymphatic system.

188 Richard Lower (1631–1691) wrote about transfusions and the cardiopulmonary system.

189 Thomas Willis (1621–1675), a founding member of the Royal Society of London.

190 Reinier de Graaf (1641–1673), a colleague of Ruysch and specialist in reproductive biology.

191 Jan Swammerdam (1637–1680), a friend and colleague of Ruysch.

192 Anton Nuck (1650–1692), a Dutch anatomist.

193 Joseph-Guichard Du Verney (1648–1730), a French anatomist.

194 Louis de Bils (1624–1671). Part gifted amateur, part charlatan, he won a reputation in Dutch scientific circles for his highly advanced preservation technique.

195 Ruysch emerged as a public intellectual in 1665, when, as a recent graduate from the University of Leiden, he demonstrated tiny valves in the lymphatic system—a fact that directly contradicted de Bils's claims.

196 "fabricae."

197 Presumably Ruysch is referring to the perfection of his wax injection technique, which allowed him to see that there were no glands at the tips of the vessels.

198 Throughout his life, Ruysch was committed to painstaking anatomical observation and was an ardent skeptic of applying theory—or "mathematical reasoning" as he derisively expressed it—to physiology.

199 Outermost layer of cells.

200 "Olecranum."

201 Uric acid crystals, a symptom of gout.

202 "humeri."

203 "ex Aethiope."

Bibliography

Anemone, Anthony. “The Monsters of Peter the Great.” *Slavic and East European Journal* 44, no. 4 (2000): 583–602.

Asma, Stephen T. *On Monsters: An Unnatural History of Our Worst Fears*. Oxford: Oxford University Press, 2012.

Asma, Stephen T. *Stuffed Animals and Pickled Heads: The Culture and Evolution of Natural History Museums*. New York: Oxford University Press, 2001.

Ball, Philip. *Curiosity: How Science Became Interested in Everything*. Chicago: University of Chicago Press, 2012.

Blancard, Stephen. *The Physical Dictionary*. 4th ed. London, 1702.

Blankaart, Steven. *The Physical Dictionary: Wherein the Terms of Anatomy, the Names and Causes of Diseases, Chyrurgical Instruments, and Their Use, Are Accurately Describ'd. … by Stephen Blancard*. London: Printed by R. B. for Sam. Crouch, and John & Benj. Sprint, 1715.

Bleker, O. P., and Jozien J. Driessen. *Geloof alleen je eigen ogen: een actuele kijk op de Anatomische preparaten van Frederik Ruysch (1638–1731)*. Hilversum: Uitgeverij Lias, 2017.

Boer, Lucas, Anna B. Radziun, and Roelof-Jan Oostra. 2017. “Frederik Ruysch (1638–1731): Historical Perspective and Contemporary Analysis of His Teratological Legacy.” *American Journal of Medical Genetics* 173A, no. 1 (2017): 16–41.

Copenhaver, Brian P. *The Book of Magic: From Antiquity to the Enlightenment*. London: Penguin, 2017.

Ebenstein, Joanna, and Colin Dickey. *The Morbid Anatomy Anthology*. Brooklyn, NY: Morbid Anatomy Press, 2015.

Goethe, Johann Wolfgang von. *Faust: A Tragedy*. Trans. Martin Greenberg. New Haven: Yale University Press, 2014.

Hansen, Julie V. “Resurrecting Death: Anatomical Art in the Cabinet of Dr. Frederik Ruysch.” *Art Bulletin* 78 (1996): 663–679.

Harrington, Anne. *Reenchanted Science: Holism in German Culture from Wilhelm II to Hitler*. Princeton: Princeton University Press, 2009.

Huistra, Hieke. *Preparations on the Move: The Leiden Anatomical Collections in the Nineteenth Century*. Leiden: Proefschrift Universiteit, 2013.

Kooijmans, Luuc. *Death Defied: The Anatomy Lessons of Frederik Ruysch*. Leiden: Brill, 2011.

Lemire, Michel. *Artistes et mortels*. Paris: Chabaud, 1990.

Luyendijk-Elshout, A. M. "Death Enlightened: A Study of Frederik Ruysch." *Journal of the American Medical Association* 212 (April 1970): 121–126.

Mazierski, David. "The Cabinet of Frederick Ruysch and the Kunstkamera of Peter the Great: Past and Present." *Journal of Biocommunication* 38 (2012): E31–E40.

Motherby, G. *A New Medical Dictionary, or General Repository of Physic*. London: J. Johnson, 1785.

Mulder, Willem J. *Frederik Ruysch Anatomical Collection in St. Petersburg: History, Storage, and Restoration*. St. Petersburg: Kunstkamera, 2004.

Purcell, Rosamond Wolff, and Stephen Jay Gould. *Finders Keepers: Eight Collectors.* London: Pimlico, 1993.

Radzjun, Anna Borisovna, Yuri Chistov, Ksenia Nosovskaya, and Julia Kupina. *The First Scientific Collections of Kunstkamera: Exhibition Guide*. St. Petersburg: Kunstkamera, 2012.

Richards, Robert J. *The Romantic Conception of Life: Science and Philosophy in the Age of Goethe*. Chicago: University of Chicago Press, 2002.

Rooy, Laurens de, and Hans van den Bogaard. *Forces of Form*. Amsterdam: Vossiuspers UvA, 2009.

Ruysch, Frederik. *Alle de ontleed- genees- en heelkundige Werken*. Amsterdam: de Janssoons van Waesberge, 1744.

Contributors

STEPHEN ASMA is Professor of Philosophy and Humanities at Columbia College Chicago, where he is a Senior Fellow of the Research Group in Mind, Science, and Culture. Asma is the author of ten books, including *The Evolution of Imagination*; *On Monsters: An Unnatural History of Our Worst Fears*; and *Stuffed Animals and Pickled Heads: The Culture and Evolution of Natural History Museums*. He writes regularly for the *New York Times* and *Aeon* magazine.

PHILIPPE COMAR is a fine artist, set designer, exhibition curator, and writer, and was professor of morphology at the Beaux-Arts in Paris from 1979 to 2019. His work has been exhibited at the Centre Pompidou, the Venice Biennale, the Mucsarnok-Kunsthalle in Budapest, and the Picasso Museum in Barcelona. He designed and produced the exhibition "Sténopé," dedicated to perspective and simulation of space, at the Cité des Sciences et de l'Industrie in Paris. In 1999 he designed the set for the ballet *Orison*, by Pierre Darde, at the Opéra National de Paris. He took part in the production of major thematic exhibitions and catalogues dealing with the body and its representation, such as *L'Âme au corps, art et science* (Grand Palais, Paris, 1993); *Identity and Alterity* (Venice Biennale, 1995), *L'Art du nu* (Bibliothèque nationale de France, 1997); *Mélancolie, génie et folie en Occident* (Grand Palais, Paris, 2005); *The 1930s: The Making of "The New Man"* (National Gallery of Canada, 2008); *Figures du corps, une leçon d'anatomie à l'École des beaux-arts* (Paris, 2010); *Lucian Freud, the Studio* (Centre Pompidou, Paris, 2010); *Crime et châtiment* (Musée d'Orsay, Paris, 2014); and *Duchamp, la peinture même* (Centre Pompidou, Paris, 2014). He is also the author of numerous books, fictions, novels, and essays, among them *Images of the Body*.

ELEANOR CROOK is a sculptor and wax modeler who works with several international medical museums. She trained in sculpture at Central St Martins and the Royal Academy Schools, studying the body from an aesthetic and a medical viewpoint in parallel and working from life and as a medical artist in the dissecting room. She is artist in residence at King's College's Gordon Museum of Pathology and has worked on projects with the Vrolik Museum Amsterdam, the Museum Dr. Guislain in Ghent, Ghent University museum, the Anatomy Museum of Cagliari in Sardinia, the Florence Nightingale Museum, the Wellcome Collection, and the Museo La Specola in Florence. Her work is in the collections of the Gordon Museum of Pathology at Guy's Hospital, the Museum of Pathology at the University of Padua, the Royal Pharmaceutical Society London, and the Hunterian Museum Royal College of Surgeons of England. Current projects include museum conservation of Ziegler and other historic wax models, a piece for the upcoming show

"Anatomy and Beyond" at the Pauls Stradiņš Museum for History of Medicine in Riga, Latvia, and a major bronze commission for the new Wellcome Galleries of the Science Museum London.

JOANNA EBENSTEIN is a Brooklyn-based artist, writer, curator, graphic designer, and independent scholar. She is the creator of the Morbid Anatomy blog, library, and event series, and was cofounder and creative director of the recently shuttered Morbid Anatomy Museum in Brooklyn. Her books include *Anatomica: The Exquisite and Unsettling Art of Human Anatomy* (with Marie Dauenheimer); *Death: A Graveside Companion*; *The Anatomical Venus*; *The Morbid Anatomy Anthology* (with Colin Dickey); and *Walter Potter's Curious World of Taxidermy* (with Pat Morris). Her work explores the intersections of art and medicine, death and culture, and the objective and subjective.

RICHARD FAULK is an author, educator, and intellectual magpie. He earned an MA in comparative literature from the University of California, Irvine, for his research on the intersection of science, magic, and political spectacle in the baroque era, work that he subsequently, and improbably, parlayed into two nonfiction books for young adults, *Gross America* and *The Next Big Thing*. He has also written on music, popular culture, medical oddities, time travel, and the science of cannabis. Recently he has been researching and lecturing on curse tablets and other forms of popular magic in antiquity, particularly as means of obtaining justice for women, slaves, the urban poor, and other marginalized classes.

LUUC KOOIJMANS is a Dutch historian and author of the only biography of Frederik Ruysch, *Death Defied: The Anatomy Lessons of Frederik Ruysch* (Leiden, 2011). He taught at the University of Amsterdam before embarking on a career as an independent writer. He was awarded the Dutch Cultural Foundation Humanities Prize for his oeuvre in 2004, and in 2008 his book *Gevaarlijke kennis* was chosen as Best History Book of the Year. His work has been translated into English, Russian, and Polish.

WILLEM J. MULDER was keeper and curator of the Anatomy, Pathology, Obstetrics, and Ophthalmology collections of the Leiden Medical Faculty (1975–1991), as well as curator of the medical collection at Utrecht University Museum (1991–2002). From 2002 to 2016 he worked on the project of restoring Ruysch specimens for the St. Petersburg Kunstkamera.

BERT VAN DE ROEMER is a university lecturer in the Cultural Studies department of the University of Amsterdam. His fields of interests are the history of collections, the relationship between the visual arts and the natural sciences in past and present, cultural life in Amsterdam in the early modern period, and modern museology, especially current museum installations. He has published on the Dutch collectors Simon Schijnvoet, Frederik Ruysch, and Levinus Vincent, and the art theorists Samuel van Hoogstraten and Willem Goeree, among other subjects. Recently he worked on cultural industries in Amsterdam in the early eighteenth century and Dutch collections of natural history during the French period.

LAURENS DE ROOY is a historian of science and curator director of Museum Vrolik, the anatomical museum of the University of Amsterdam, located in the Academic Medical Center. He received his PhD in 2009 for a study of Amsterdam anatomist Lodewijk Bolk (1866–1930). He also published the book *Forces of Form* (2009) about the Museum Vrolik. In 2012 De Rooy refurbished the permanent exhibition of the museum, with more emphasis of the historical context of the collection. De Rooy also lectures in medical history. His current interest is in the history of physical anthropology.

Acknowledgments

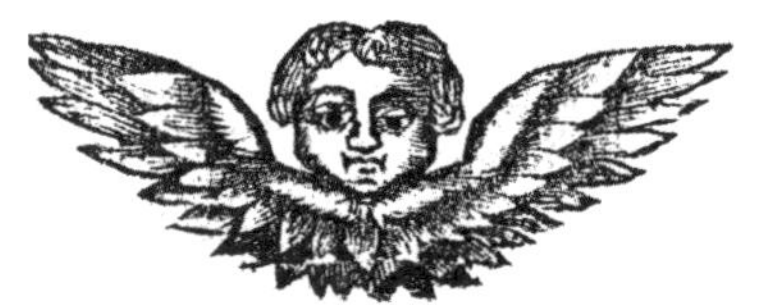

DELUXE EDITION SUPPORTERS

The editor would like to offer sincere and heartfelt thanks to those who supported this project by purchasing deluxe editions, which enabled us to use high-quality paper and four-color printing throughout.

Those individuals and institutions are:

ARCHIVES & SPECIAL COLLECTIONS, AUGUSTUS C. LONG HEALTH SCIENCES LIBRARY, COLUMBIA UNIVERSITY

SHERWIN BORSUK MD

THE CENTER FOR THE HISTORY OF MEDICINE AT BERNARD BECKER MEDICAL LIBRARY

MARIE DAUENHEIMER, MA, CMI, FAMI

DANNY ELFMAN

JAMIE HENDERSON, FORSAKEN INK

MELISSA KARABAN

ROBYN MEWSHAW

AUDREY NIFFENEGGER

AME SIMON
"In loving memory of my late husband Marc Carter, and to Morbid Anatomy for reminding me that in death our souls begin their next adventure into the mystery of life and that there is nothing to fear."

THE VIRGINIA FOX STERN CENTER FOR THE HISTORY OF THE BOOK IN THE RENAISSANCE, JOHNS HOPKINS UNIVERSITY

GEORGINA VOEGELE

DR. ANDREW ZBAR

OTHER THANKS

For support both moral and material, I would like to thank my husband, Bryan Melillo, and my good friend Eleanor Crook (without which this book would be a very different beast indeed!). I would also like to thank the many friends and colleagues who lent varieties of inspiration and assistance: Amy Slonaker, Andries J. van Dam, Bart Grob, Carl Schoonover, Catherine Crawford, Elena Osadchaia, Eric Huang, Evan Michelson, John Troyer, Laetitia Barbier, Laurence de Rooy, Macabee Montandon, Marie Dauenheimer, Mark Pilkington, Matt Murphy, Patricia Llosa, Rebecca Messbarger, Dr. Reina de Raat, Dr. Ronald L.A.W. Bleys MD, Rosamund Purcell, and Ross MacFarlane, and William Baker. Thanks to Spencer Lamm for reading early drafts of the introdution.

I would also like to give thanks to my family: first, to my Oma and Opa, Jewish refugees from Hitler's Vienna, doctors who loved art and history, who taught me to value the same; I think they would appreciate this book, and the enigmatic man whose work is its subject. Thanks also to my parents, Sandy and Robert Ebenstein; my stepmother, Judy Ebenstein; my aunt and uncle, Judy and Dick Grose; my sisters, Donna and Laura Ebenstein, my brother-in-law Jove Graham, and my niece and nephew, Clara Jean and Thomas Benno Graham.

I would also like to thank all of the institutions that so generously allowed us to include images from their collections in these pages. First and foremost to Elisabeth Brander of the Bernard Becker Medical Library, Washington University in St. Louis School of Medicine, who offered us complete use of her collection, and assisted in so many ways. I am deeply grateful for the ability to include photos of the original specimens, courtesy of the Peter the Great Museum of Anthropology and Ethnography (Kunstkamera) of the Russian Academy of Sciences; Museum Bleulandinum, University Medical Center Utrecht in The Netherlands; and Leiden University Medical Centre (LUMC), also in The Netherlands. Thanks also to the Amsterdam Museum for allowing us to include a wonderful portrait of Ruysch and, as always, to Wellcome Images and The Wellcome Library for making availble their incredible collection of high-resolution images free of charge.

I would also like to thank all the contributors for their wonderful work. Many thanks are due first to Richard Faulk, a long-time friend and gifted translator, who used his keen aesthetic and historical eye to pinpoint the most important parts of Ruysch's book, and whose sparkling tranlation brought warmth and life into what could have been a very dry text. Special thanks to Eleanor

Crook, who contributed not only an essay but also brilliant illustrations. And thanks to all the other essayists: Bert van de Roemer, Laurens de Rooy, Luuc Kooijmans, Philippe Comar, Stephen Asma, and Willem J. Mulder.

Finally, I woud like to thank the excellent editorial team at the MIT Press that made this book a reality. First off, Matthew Browne, who saw the possibility in this project, and whose patient and thoughtful shepherding saw it through to its enhanced completion. Thanks also to Janet E. Rossi, director of EDP and Publishing Logistics, to Matthew Abbate, who copyedited the manuscript, and to Erin Hasley, who designed the book's cover.

Index

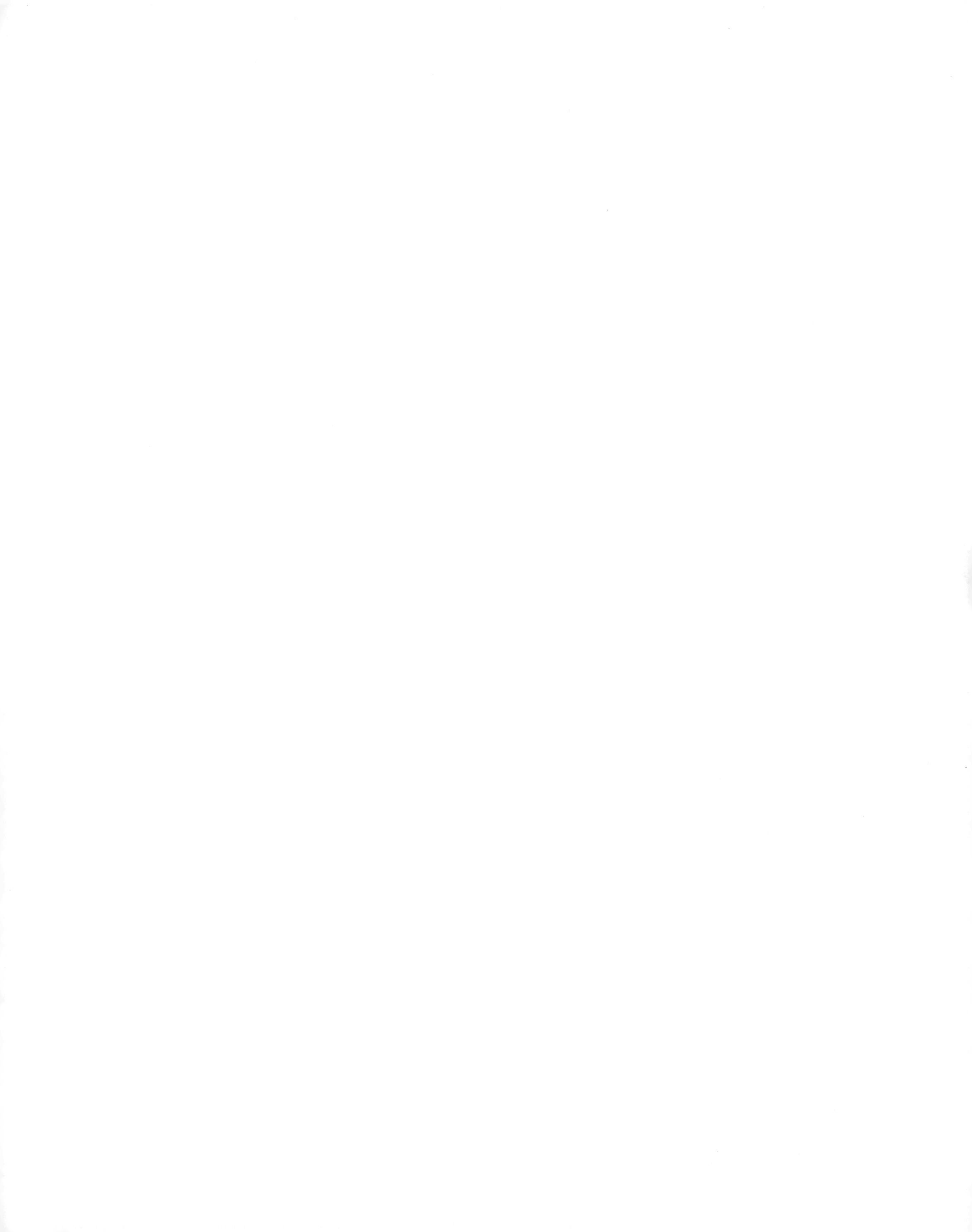

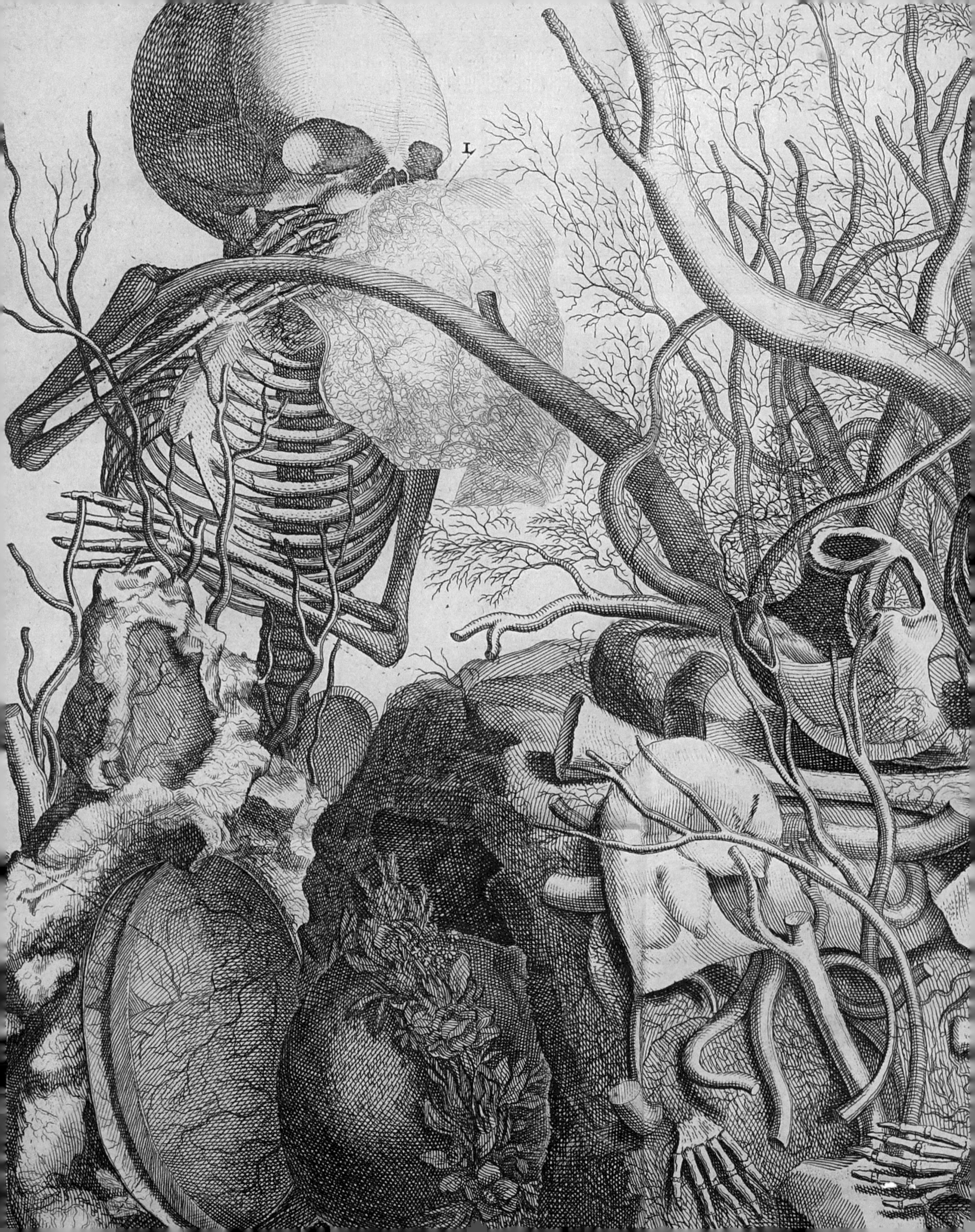
L